Plants of
Central Texas
Wetlands

Plants of Central Texas Wetlands

Scott B. Fleenor

and

Stephen Welton Taber

Texas Tech University Press

Copyright © 2009 by Texas Tech University Press

This book is typeset in Adobe Warnock Pro. The paper used in this book meets the minimum requirements of ANSI/NISO Z39.48-1992 (R1997).
∞

Designed by Kaelin Chappell Broaddus

Library of Congress Cataloging-in-Publication Data
Fleenor, Scott B., 1962–
   Plants of Central Texas wetlands / Scott B. Fleenor and Stephen Welton Taber.
      p. cm.—(Grover E. Murray studies in the American Southwest)
      Summary: "Describes the plants of the Ottine Wetlands of south central Texas, within and surrounding Palmetto State Park. This important ecological region has been little studied and has not been fully described previously. Includes an introduction to the wetlands, descriptions of the plants, color plates, a complete checklist, and a glossary"—provided by the publisher.
      Includes bibliographical reference and index.
      ISBN 978-0-89672-639-0 (cloth bound : alk. paper) 1. Wetland plants—Texas. I. Taber, Stephen Welton, 1956– II. Title. III. Series.
      QK188.F58   2009
      581.7'6809764—dc22
                          2008035780

Printed in Korea
09 10 11 12 13 14 15 16 17 / 9 8 7 6 5 4 3 2 1

Texas Tech University Press
Box 41037
Lubbock, Texas 79409-1037 USA
800.832.4042
ttup@ttu.edu
www.ttup.ttu.edu

Dedicated to the Memory of Harvey Soefje
(August 24, 1932–January 1, 2002)

Contents

Preface

The Ottine Wetlands of Central Texas, or more precisely, south-central Texas, lie on both public and private properties, so we thank Sheriff Glen A. Sachtleben and Mr. Garry Henderson of the Gonzales County Appraisal District for guiding us to private landowners when we began our study. Permission to enter and explore was kindly granted by Evelyn Pettus, Will Soefje, and the late Harvey Soefje, to whom this book is dedicated for graciously allowing us free rein in Soefje Swamp. We thank Corinna Walker and Dale Walker for introducing us to Harvey Soefje in the field.

Access to the publicly owned wetlands requires no permission because these lie within the boundaries of Palmetto State Park. Nevertheless, a permit is required to collect plants and animals for purposes of identification, and thus we thank David Riskind of the Texas Parks and Wildlife Department for providing Scientific Study Permit No. 21-01. Park Superintendent Mark Abolafia-Rosenzweig alerted us to the presence of local species that might otherwise have been overlooked; apprised us of relevant developments, including flowering periods, rainfall, temperature, and flood status; and provided answers to many other questions. Dr. Billie Turner of the University of Texas contributed a copy of his two-volume *Atlas of the Vascular Plants of Texas,* which was published just in time for use in the present work. Jason Singhurst, botanist with the Texas Parks and Wildlife Department Wildlife Diversity Program, apprised us of the known occurrence of *Sphagnum* mosses in Gonzales County during the later stages of manuscript revision.

Special thanks are extended to Director Noel Parsons of Texas Tech University Press for encouraging eventual publication of these results long before a manuscript was ready for submission. Dr. James Dixon of Texas A&M University kindly gave permission to use a map that formed the basis of Map 1-1, and Ms. Cheryl O'Brien and Dr. David Riskind alerted us to the fact that U.S. Geological Survey and Texas State Park maps lie in the public domain. Dr. Riskind also read the manuscript in its early stages and provided helpful criticism.

We close these acknowledgments with homage to the work done by two field biologists of days gone by. In the early years of the twentieth century, the botanist Edwin Robert Bogusch conducted a survey of plants in an area near our study sites but not seen by us (Bogusch 1928, 1930). This was followed in the middle of the century by Gerald G. Raun's zoological survey of mammals, reptiles, and amphibians (Raun 1958, 1959). Raun regretted that a study of his favored subjects had not been undertaken by Bogusch, and Bogusch might well have appreciated the same of Raun when the latter's study began. If each had surveyed both plants and animals instead of only one or the other, we would already have some basis for an evaluation of historical changes in flora and fauna that might, in turn, reflect changes in the water relations or "hydrology" of the area. However, even had they done so, the study sites of these two pioneers were some distance apart, and this would have been a complication for any comparisons between yesterday and today. We chose the previously unstudied invertebrates (Taber and Fleenor 2005) and the plants that are so integral a part of any wetlands for our own emphasis.

The relevance of our study was underscored by a related regional wetland overview and history that somehow overlooked the work of Bogusch, Raun, and others and that appeared while our fieldwork was just getting under way (MacRoberts and MacRoberts 2001). The West Gulf Coastal Plain is defined therein as a region bounded on the east by the Mississippi River from southern Illinois to the Gulf of Mexico and on the west by a more undulatory line extending from Fort Smith, Arkansas, in the north to Uvalde, Texas, in the extreme south. Thus, the Ottine Wetlands lie near the western edge of a region equal in size to about two-thirds the area of Texas, but they are not mentioned in the overview of that region's bog communities, which must by definition include the marshes in the south-central portion of the state.

The overview would have classified our marshlands as "muck bogs" had they received mention, though by the definition we employed there are no true bogs at all in Texas (see Glossary). MacRoberts and MacRoberts acknowledged that only the eastern portion of the defined region has been emphasized, a fact evident in their figure 11 (p. 28), which confines all the West Gulf Coastal Plain wetlands to a single saddle-shaped enclosure in eastern Texas and western Louisiana, despite the existence of the Ottine Wetlands about 160 km southeast of the saddle. The authors of that overview did point out the complete lack of any invertebrate survey for any bog community, and certainly that was the case within the entire West Gulf Coastal Plain. We filled that gap with our previous volume, and here we do the same for the overlooked flora of the Ottine Wetlands. It is also true that, of the three bog categories recognized in the summary by MacRoberts and MacRoberts, muck bogs feature the greatest percentage of obligate wetland species.

According to that treatment great differences would be expected in the species compositions of muck bogs inside and outside the saddle. For example, the carnivorous pitcher plant *Sarracenia alata* A. Wood was the indicator species that defined the geographic distribution of their bog communities in Texas and Louisiana. However, no pitcher plants occur in the marshes (muck bogs) of the Ottine Wetlands, though they are home to ferns and sedges also recognized as indicator species for these same habitats (MacRoberts and MacRoberts 2001, 45). Furthermore, though the authors found that nearly all the bogs known to them were free of exotic wetland-requiring plants, we encountered three such species and list them in our own Appendix 2. They are the very conspicuous yellow flag iris (*Iris pseudacorus*) and water-hyacinth (*Eichhornia crassipes*), and the less obvious umbrella sedge (*Cyperus involucratus*). Appendix Two of MacRoberts and Mac-Roberts is a list of plants occurring in the muck bogs, wetland pine savannas, and hillside bogs of West Gulf Coastal Plain bog communities as the authors knew them, and surprisingly, this list shows little overlap with the apparently unique association of species in the muck bogs or marshes of the Ottine Wetlands.

Finally, during the preparation of this book, a reviewer suggested that we describe how we went about our work. First, we consulted the scientific literature and topographic maps to see what was known about the flora and fauna and where the wetlands were located, respectively. Second, we obtained a Scientific Study Permit from the Texas Parks and Wildlife Department. Such a permit is required by anyone who collects plants and animals in state parks. Third, while making weekly trips to the public land of Palmetto State Park, where no permission to enter is required, we sent e-mails and letters and made phone calls to those who owned the private swamps and marshes surrounding the park. Even finding these names was a form of exploration, for we found it necessary to consult state and county officials, although we had almost no information to go on besides the location of the land itself. Most of the time we reached the right parties; whenever we did, they granted us permission to enter, observe, and collect. We were pleasantly surprised by this courtesy, and the following cannot be emphasized too strongly: **Never trespass on private property.** It is the responsibility of everyone who takes to the field to be aware of his or her surroundings, not only the dangers presented by plants, animals, and the deep muck of the wetland itself but also the ownership of those very lands. The authors know of cases in which local landowners "escorted" trespassers off their property, and in similar scenarios far from these wetlands, reactions were much, much worse.

With verbal permission for private land, and with a written scientific study permit for public land, we began photographing, collecting, and identifying as many plants and animals without backbones as we could. When the library of the University of Texas at Austin did not have the relevant field guides on its shelves, we attempted to obtain them from other institutions through Interlibrary Loan.

Plants of
Central Texas
Wetlands

1

The Ottine Wetlands of Central Texas

The Ottine Wetlands comprise a relict ecosystem that lies on the floodplain of the San Marcos River near the junction of Guadalupe, Caldwell, and Gonzales counties in south-central Texas (Maps 1-1, 1-2). Here, after entering Gonzales County, the southeasterly meandering river bisects a northeast-southwest trending ridge of sand and rock over 30 m in height. Eight kilometers southeast of the town of Luling (Caldwell Co.) it is deeply entrenched, flowing nearly 30 m below the nearby village of Ottine (Gonzales Co.), for which the wetlands are named. They are unique for their resemblance to river bottom forests and swamps of the Big Thicket (Raun 1958; Nixon, Chambless, and Malloy 1973; Mohlenbrock 2002) and to marshes of the Texas coast (Nixon 1963), from which they are separated by 160 km or more. Indeed, these wetland ecosystems constitute a lowland counterpart to the Lost Pines of Bastrop, Fayette, and Caldwell counties (Taber and Fleenor 2003; Fleenor and Taber, *Plants of the Texas Lost Pines*, forthcoming), 56 km to the north and east. Notable too is their location astride the 98th meridian of longitude, the biogeographic divide separating the eastern plants and animals of the United States from its western flora and fauna and for the diverse mixture of species that occurs there. The same holds for the north-south direction, as many plants and animals of subtropical or even Neotropical affinities, some occurring as far south of the equator as Argentina, reach their northernmost limits here or nearby. However, most of the biota occurring here are of eastern and Nearctic origin.

Biotic Regions

According to the standard reference by Blair (1950), Texas is divided into seven biotic zones or provinces, and the Ottine Wetlands are contained within the post oak woodland/savannah of the namesake "Texan" province. Immediately to the east of this zone is the Austroriparian province, representing an extension of the flora and fauna of the southeastern United States into the state. Central Texas wetlands are more strongly affiliated with this zone than any other. To the west, but ironically even closer to the wetlands than the Austroriparian, lies the Balconian

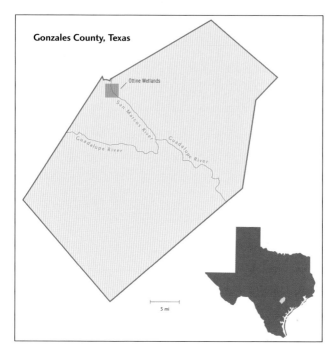

Map 1-1
The Ottine Wetlands lie along the San Marcos River in Gonzales County, Texas (river width not to scale).

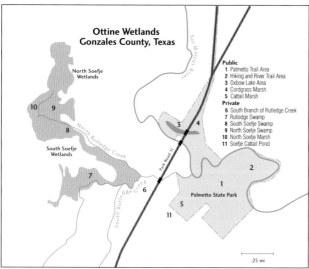

Map 1-2
The Ottine Wetlands are both privately and publicly owned.

province, characterized by prairie and southwestern elements. To the south lies the Tamaulipan zone, home to subtropical Mexican and even Neotropical species that reach their northern limits here. These four biotic zones (Texan, Austroriparian, Balconian, and Tamaulipan) correspond to the Appalachian (Interior Eastern), Southeastern Coastal Plain, Prairie, and Tamaulipan floristic provinces

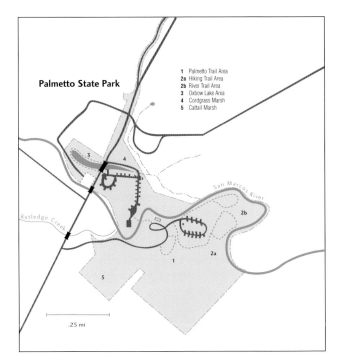

1 Palmetto Trail Area
2a Hiking Trail Area
2b River Trail Area
3 Oxbow Lake Area
4 Cordgrass Marsh
5 Cattail Marsh

Palmetto State Park

San Marcos River

Rutledge Creek

.25 mi

Map 1-3
Publicly owned
wetlands lie within
Palmetto State Park.

(Thorne 1993), respectively, and the following respective vegetation types: Post Oak Savannah and Blackland Prairie, East Texas Pineywoods, Hill Country Savannah (Edwards Plateau), and South Texas Brushland. The Ottine Wetlands thus appear as disjunct Austroriparian outliers near the present-day junction of the Texan, Balconian, and Tamaulipan regions, with their associated floristic provinces and vegetation types. It is here that the wetland flora and fauna of the southeastern United States reach their western limits.

History of Human Interest

As of this writing there had been just over seventy-five years of research and publications dealing specifically with the Ottine Wetlands. The earliest work was that of the botanist Bogusch (1928). Palmetto State Park (Map 1-3) was created soon afterward in 1933 (Shearer 1956; Maxwell 1970), although the apparent connection is not clear. The brief flurry of publications that followed gives some hint of the interest in these wetlands that supported the inclusion of some of them in a state park. This early concern for the protection of the swamps and marshes may be explained by their unexpected occurrence so far west of their Austroriparian counterparts and by the surprisingly diverse mix of eastern, western, and southern plants and animals occurring here in unique associations found nowhere else in the world. Research, reports, and efforts to excite public interest grew to include checklists of birds (Kirn 1935), amphibians and reptiles (Parks 1935b), and butterflies (Parks 1935c) while expanding upon the inaugural botanizing of Bogusch (Parks 1935a; Tharp 1935). Older literature such as this is sometimes more valuable than recent work because it allows comparisons of historical changes in habitats, hydrology, and more specifically, the compositions of their flora and fauna. For a history of the area written at the time of Palmetto State Park's creation, see Hildebrand (1935). Recent bird checklists are those of Hartigan and Lasley (1987) and Rogers (1999). For the early work on mammals, reptiles, and amphibians, see Raun (1958, 1959).

Glacial History

Even plants that died long ago became objects of interest—and not entirely academic interest, either. Numerous deposits of partly decomposed plant material known as peat occur throughout the area. Several of these deposits were cored, sampled, and analyzed to determine their depth, extent, and quality with a view to mining and sale as a soil conditioner (Chelf 1941). Though they were never brought into commercial production, mining of peat from several "bogs" to the north and east had become a small local industry by the late 1930s (Plummer 1941, 1945; Chelf 1941). Peat deposits up to 15 m deep were reported in the Ottine area. Two additional peatlands of half a hectare or less in extent and with peat accumulations of less than 1 m were reported to the west in Guadalupe County, where they were associated with springs and a small tributary of the Guadalupe River (Chelf 1941). These are the southwesternmost known peatlands in the eastern United States. Pollen analyses of cores taken from marshes and "bogs" indicate that the wetlands are at least twelve thousand years old and that their floral composition has been changing with the drying climate since the end of the last glacial advance eighteen thousand years ago, though there is disagreement as to the extent of the change (Graham 1958; Graham and Heimsch 1960; Patty 1968; Larson, Bryant, and Patty 1972). For example, some of the first pollen analyses of cores from Texas peat deposits to the north and east indicated the presence of spruce (*Picea* sp.) and fir (*Abies* sp.) (Potzger and Tharp 1943, 1947, 1954) in the region, but pollen analyses of peat from the Ottine area have not corroborated this, and it remains a subject of some dispute (Patty 1968; Bryant 1977). However, the presence of smooth alder (*Alnus serrulata* (J. Dryander *ex* W. Aiton) C. von Willdenow), river birch (*Betula nigra* C. Linnaeus), and possibly sweetgum (*Liquidambar styraciflua* C. Linnaeus) here in the last ten thousand years is well supported (Patty 1968; Larson, Bryant, and Patty 1972). All these presently occur farther east in the botanically distinctive wetlands and river bottoms of East Texas. Blackgum (*Nyssa sylvatica* H. Marshall) occurs in wetlands as far west as Brazos County, whereas its close relative water tupelo (*N. aquatica* C. Linnaeus) only reaches into extreme East Texas near the Louisiana border. Neither occurs as far west as the Ottine Wetlands today.

However, to say that the Ottine Wetlands remain poorly studied even after seventy-five years remains an understatement. Floristic surveys of more eastern Texas wetlands have been conducted recently (Rowell 1949; Nixon, Chambless,

Fig. 1-1
The San Marcos River flowing near the entrance to Palmetto State Park.

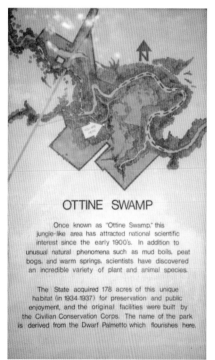

Fig. 1-2
Ottine Swamp as portrayed on the stone tower in Palmetto State Park.

OTTINE SWAMP

Once known as "Ottine Swamp," this jungle-like area has attracted national scientific interest since the early 1900's. In addition to unusual natural phenomena such as mud boils, peat bogs, and warm springs, scientists have discovered an incredible variety of plant and animal species.

The State acquired 178 acres of this unique habitat (in 1934-1937) for preservation and public enjoyment, and the original facilities were built by the Civilian Conservation Corps. The name of the park is derived from the Dwarf Palmetto which flourishes here.

and Malloy 1973; MacRoberts and MacRoberts 1998), but no such work has been published for the Ottine Wetlands since the initial publications of Bogusch (1928), Tharp (1935), and Parks (1935a). The work of McAllister, Hogland, and Whitehouse (1930), Whitehouse and McAllister (1954), and Ellison (1964) on the liverworts of Texas, and that of Whitehouse (1955) on the mosses of Texas and Lodwick and Snider (1980) on Texas *sphagnums*, contained numerous valuable collection records from the Ottine Wetlands. Thanks to David Riskind of the Texas Parks and Wildlife Natural Resources Division we also discovered, filed in the state's archives, the unpublished Natural Area Survey conducted by Williams and Watson in 1978. This served to corroborate and amplify our own work. The lack of research is remarkable considering the proximity of the site to several major universities that lie, in turn, near the center of a large, agriculturally oriented state. We set out to revisit the localities, many of them on private land, of these earlier botanists and conduct a modern survey of our own with a view to documenting some of the remarkable plants found here and the floral changes that may have occurred during the intervening seventy years. One landowner, while pointing in the direction we sought, described our destination as "wilderness" that had not been visited in years. Another, a former schoolteacher, recalled the field trips of botany classes led by Dr. Benjamin Carroll Tharp and others of the University of Texas during the 1930s, 1940s, and 1950s.

We were unable to examine two small areas that we wished to investigate despite several attempts to contact the landowners by letter and telephone. A lack of response is understandable, because the findings of biodiversity specialists often cause private landowners to regret their kindness at a later date. One landowner remarked to us, perhaps only half joking, "Now don't find any endangered species!" The first of these sites lies near the base of Red Hill Overlook, just inside the northern entrance to Palmetto State Park. Its ash swamps and *Sphagnum* marshes were the first of the Ottine Wetlands to be studied (Bogusch 1928, 1930). Strangely, there is no indication that Bogusch was aware of the wetlands encompassed by the soon-to-be-established park, although they are only a few miles south of his own site.

Recent History of the Wetlands

By the middle of the twentieth century, the swamps and marshes that Bogusch studied had diminished drastically, were unrecognizable as such, and were perhaps eventually destroyed (Raun 1958). Thus, there may have been little for us to look at below Red Hill Overlook. Some of this change occurred within a decade of that botanist's fieldwork. At the time of our own visits we were unable to ascertain visually the situation, even from the fence line at the base of Red Hill Overlook, because of intervening brush and trees. Tangles of trunks and denuded branches in winter were nearly as great an interference as the prolific leaves of spring and summer.

Botanists had found peat moss growing at this site, and its presence was remarkable as defining the southwestern limit of the genus *Sphagnum*'s range in the eastern United States. Later records of it came from the Soefje wetlands and from Hershop "bog" (Chelf 1941), the latter locality being the second of the two sites that we wished to explore but could not obtain permission to do so. That site is, or at least was, better described as a marsh than a bog and lies a short distance west of Palmetto State Park. Chelf reported that the extinction of *Sphagnum* there had occurred within the last thirty years. Later, the authority on this small peatland summarized its condition as "dead" and undergoing erosion and silting (Patty

Fig. 1-3
The view from Red Hill
Overlook.

Fig. 1-4
Peatland formerly
described as a bog.
The herbaceous
plants in the center
of the picture are soft
rushes (*Juncus effusus*),
and just behind are
southern wax-myrtles
(*Myrica cerifera*).

1968). We devoted much time to the search for *Sphagnum* on both private and public properties, but we never saw any of the three species previously reported (*S. imbricatum, S. palustre,* and *S. subsecundum*) despite a report of success only thirty years before (Lodwick and Snider 1980). It appears that we walked over the same ground described by those authors. Recently, however, several *Sphagnum* sites have been documented in Bastrop and Gonzales counties, including its rediscovery in the Soefje wetlands (Jason Singhurst, pers. comm. 2007, 2008).

Geology

Concomitant with the peatland and palynological work were studies of the underlying geology and hydrology that make life possible for eastern plants and animals requiring abundant moisture to persist this far west. The west-facing escarpment through which the meandering San Marcos cuts, and that flanks the river on either side of its broad floodplain, is composed of the Carrizo and lower Reklaw

(Newby member) sandstone formations of Eocene age (about 50 million years old). The thick, soft, orange-yellow Carrizo weathers to a fine, pale, buff-colored sand and is capped near the top of the ridge by the coarse, hard, dark red sandstone and conglomerate ledge that marks the lower portion (Newby sandstone member) of the Reklaw Formation. Near their contact may be seen numerous nodular and potsherdlike eroded remnants of concretions resulting from the percolation and subsequent precipitation of iron- and sulfur-charged waters into the Carrizo from the overlying Reklaw. The Carrizo aquifer is composed of groundwater held in these sands, which are bracketed above and below by relatively impermeable strata (the upper Reklaw [Marquez shale member] and the clayey Calvert Bluff Formation of the Wilcox group, respectively). Local faulting has in some cases changed the lateral juxtaposition of these layers and exposed permeable, water-bearing strata at the surface, creating seeps and springs. In addition, the broad floodplain terraces of the San Marcos are composed of very fine silts and clays transported from the Blackland Prairie and Edwards Plateau (via the Blanco River) and deposited during flood stages. These sediments impede drainage following flooding. Where these relatively impermeable floodplain sediments abut sandy uplands along valley margins, springs and seepage areas often occur. The erosional history of the river itself is preserved on the floodplain surface in the form of abandoned channels and cutoff meanders that may later become swamps, sloughs, and oxbow lakes, often lying many feet above the present-day level of the river. These remain wetlands through a combination of impeded drainage, seepage from surrounding uplands, and replenishment during floods. Floodplain erosion during floods may also create or maintain channel-like depressions and ponds that sometimes pockmark the floodplain surface. The Ottine Wetlands feature examples of all of these. The geology and hydrology of this region have been detailed by Cumley (1931), Bullard (1935), and King (1961).

Soils

The soils of the Ottine Wetlands differ markedly from those of the surrounding uplands. This is a result of flooding frequency and degree of saturation as well as the parent materials on or in which they develop and the consequent vegetation they support. Soils of the less frequently flooded (less than annually) upper floodplain of the San Marcos River valley develop in very fine dark brown to black silts and clays transported from the Blackland Prairie and limestone Hill Country to the west. During periods of intermittent drying they may shrink and develop deep vertical cracks and are called Vertisols in reference to this behavior. When rewetted, they swell, the cracks fill with water and close, and the soil compacts over time into a tight, sticky, gumlike texture that impedes drainage when wet. The natural vegetation type occupying these soils is Blackland Prairie. The presence of grassland vegetation may transform such soils into Mollisols through the organic enrichment of the upper layer by the decomposition of their roots and thatch, creating what is known as a mollic epipedon or prairie sod. Much of the private pasture and agricultural land in the San Marcos River valley is of this soil type. Circumneutral in pH, Mollisols have a high nutrient (base cation) content. Intermittent wetlands may occur in the form of sloughs, oxbow lakes, and ponds or "lagoons." However, these wetlands, unless they have a perennial source of groundwater (either natural or artificial), usually dry up during summer droughts.

Soils of the lower river terraces are more frequently flooded by the San Marcos River (at least annually, usually more often) with frequent deposition/erosion of silt and clay alluvium. For this reason there is little time for the development of well-defined soil horizons. Though they may show a series of subsequently buried, leached surfaces, they have no significant horizontal structure (horizon development) at depth. Known as Entisols, they are thus permanently young. In addition, they may be distinguished from Vertisols by the lack of wide (less than 1 cm), deep cracks upon drying and from Mollisols by the lack of a mollic epipedon (Soil Survey Staff 1975). Because of frequent flooding by the circumneutral waters of the San Marcos and the basic nature of their parent materials, they are nutrient rich and of neutral pH. The natural vegetation of these soils is bottomland forest or swamp. Most of the soils of the floodplain wetlands (swamps, ponds, and marshes) are of this type. Still, without a permanent source of groundwater these likewise usually dry up during the hot summers.

The soils of wetlands abutting the sandy uplands along the margins of the San Marcos floodplain differ in having a permanent source of groundwater in the form of springs and seeps. This water differs from that of the river in being acidic and laden with dissolved iron and sulfur compounds leached from the overlying Reklaw Formation. These wetlands have formed on a base of weathered Carrizo sand washed down from the neighboring uplands, sometimes as beds interlayered with silt or peat and resting at depth on fine, relatively impermeable silty or clayey floodplain material that acts to impede drainage. In these perennial wetlands plant production exceeds decomposition, and a thick organic layer of peat (a histic epipedon) develops over time, aided by acidic, anoxic, waterlogged conditions that hinder the activity of microorganisms. This consists of the partly decomposed remains of algae and the roots, stems, leaves, and propagules of mosses, ferns, sedges, grasses, forbs, shrubs, and trees. The peat is typically brown to black in color, but the secondary precipitation of iron and sulfur may give it a reddish or yellowish tint. Peat may occur in various stages of decomposition and compaction, ranging in texture and color from sedgy and fibrous brown to fine, charcoal-like dark brown or black and finely compacted to watery muck. The top layers are often knit together by the roots of living vegetation, and the entire saturated peat body may quake or undulate when jumped upon. Sometimes where water has accumulated at depth, this upper layer is actually floating on a layer of muck a meter or more deep. Wind-thrown trees are a common sight in swamp and hillside peatlands, as the waterlogged, anoxic conditions at depth discourage the growth of deep, anchoring roots. Such soils are known as Histosols.

Some of these wetland soils have a history of grassland vegetation as indicated by the presence of a dark-colored mollic epipedon (Ramsey and Bade 1977). They have subsequently become subject to flooding of varying frequency by the migrating meanders of the San Marcos River. This has subjected them to varying degrees of silting (deposition) and erosion as well as saturation ranging from intermittent to permanent. This, in turn, has the effect of transforming them in the direction of, or partially into, other soil orders, namely Entisols, Vertisols, and Histosols. Increased flooding frequency and consequent deposition of silt and clay transform them toward Entisols and Vertisols; increased saturation (waterlogging), toward Histosols. The reverse may also occur as the river migrates away or downcuts, leaving former floodplain terraces higher and drier, with Entisols developing over time into Vertisols and Mollisols. True Entisols (sandy to clayey) occur in very frequently flooded areas near the river and tributary streams, and Histosols have developed de novo in the areas of springs and seeps.

Climate

The climate of the Ottine Wetlands region may be generally described as humid subtropical with hot summers (Bureau of Business Research and Natural Fibers Information Center 1987). Annual rainfall distribution is bimodal, with peaks occurring during the late spring and early summer (April, May, and June) and again in fall (September and October). During each of these months, rainfall averages about 10 cm. As might be expected, average gauge levels of the San Marcos River correspond with these precipitation peaks. Flood levels of 9 m or more above the normal river level at Ottine (about 87 m) are not uncommon and usually occur at least once or twice per year. This (elevation 96 m) marks the level of the major floodplain terrace along the river here. The remaining months of the year average about 5 cm per month for an annual precipitation total of about 88 cm. Winter (December, January, and February) rainfall results from the interaction of cold polar fronts and moist maritime air from the Gulf of Mexico, usually occurring as light rain or drizzle, sometimes of prolonged duration. Thunderstorms account for most of the rainfall during the late spring/early summer peak with tropical storms sometimes contributing in late summer and fall. At these times moist maritime air from the Gulf and high temperatures predominate. Mid- and late summer (July and August) are usually characterized by drought conditions of high temperature (often exceeding 37.7°C) and evaporation combined with low rainfall. Temperature and evaporation peak during the month of August. A gradual amelioration of these extremes accompanied by welcome rain occurs in fall (September, October, and November).

Wetland Classification

Wetlands are difficult to classify in a manner that satisfies everyone (Lewis 2001). In fact, the word *wetland* itself is subject to competing and complicated definitions. A good one is the straightforward version offered by the U.S. Fish and Wildlife Service, which views wetlands as "lowlands covered with shallow and sometimes temporary or intermittent waters. They are referred to by such names as marshes, swamps, bogs, wet meadows, potholes, sloughs and river-overflow lands. Shallow lakes and ponds, usually with emergent vegetation as a conspicuous feature, are included in the definition, but the permanent waters of streams, reservoirs, and deep lakes are not included. Neither are water areas that are so temporary as to have little or no effect on the development of moist-soil vegetation" (Mitsch and Gosselink 2000, 29). This concept emphasizes the presence of plants adapted to at least periodically saturated soil rather than the nature of the soil itself. However, the role of the permanent Ottine waterways, such as the San Marcos River, remain paradoxically uncertain under this interpretation except for those lands inundated by its periodic floods.

We identified seven of the nine wetland types or "species" specifically listed in the U.S. Fish and Wildlife Service definition. These are marshes, swamps, wet meadows, sloughs, river-overflow lands, shallow lakes, and shallow ponds (Mitsch and Gosselink 2000). It is tempting to exhaust the inventory by classifying some of the sites as potholes and bogs, but the Ottine area is too far south to embrace such features in the strict sense. For example, the curious bowl-shaped "lagoons" of Palmetto State Park could be described as potholes, but the latter are understood to be prairie features formed by Pleistocene glaciers that never extended as far south as Texas. Instead, the lagoons are believed to be natural, possibly erosional depressions that were artificially modified during the construction of the park. Likewise, despite the fact that others have used the term *bog* for some of the habitats we visited, the true bogs of more boreal climes are fed exclusively by precipitation rather than by the additional sources of groundwater and surface runoff or drainage that are vital to all of the Ottine Wetlands (Vitt 1994, 2000) (see Glossary for wetland definitions and distinctions). *Sphagnum* moss is also a bog indicator, as is the development of peat, yet neither is as reliable as the hydrological criterion. *Sphagnum* once occurred here and has recently been redocumented here and elsewhere in Gonzales County (Jason Singhurst, pers. comm. 2007, 2008). Peat accumulation occurs in several types of wetlands as well as bogs. Most interesting are the domed areas 3.6–4.6 m across and raised up to 0.3 m above the surrounding peatland margin that were reported for Soefje and Hershop peatlands (Chelf 1941; Patty 1968). This is due to the accumulation of partly decomposed plant materials above the level of the local groundwater table as a result of impeded drainage caused by the surface vegetation and the underlying peat itself. This is a characteristic feature of true bogs, and a continuation of this development under higher rainfall conditions could conceivably lead to a true ombrotrophic (rain-fed) condition. However, because of the frequency of summer drought and the consequent slight development of doming, capillary conduction of groundwater and surface drainage were probably necessary for the persistence of these features here. As with the identification of plants and animals, it is important to keep distinctions among the names of different species clear, as the risk of misidentification has consequent ripple effects on the future work undertaken by others (Vitt 1994). We settled on a wetland classification scheme recognizing only four categories in the Ottine area: marsh, swamp, oxbow lake, and pond. Finally, we note that wetlands may be permanent or intermittent, and they may be natural, artificial, or a combination thereof. Artificial or created varieties may be intentional reclamations from previously drained natural sites, wetlands developed de novo from drier uplands, or accidental or unintended creations.

Ottine Wetland-Type Localities

In our survey work we visited eleven localities that we classified as wetlands of various types. Five of these occur on public land and six on private land. The total area of all eleven combined lies somewhere between 2.6 and 5.2 km^2. The following list shows these eleven areas of the Ottine Wetlands, annotated with their type classifications and relevant characteristics (Figs. 1-5–1-42).

Public Wetlands

1. Palmetto Trail area (elevation 97 m) (swamp, natural/artificial, potentially permanent; Figs. 1-5–1-18)
2. Hiking and River Trail areas (elevation 94.5–97.5 m) (swamp, natural, intermittent; Figs. 1-19–1-21)
3. Oxbow Lake area (elevation <96 m) (oxbow lake, natural/artificial, potentially permanent; Figs. 1-22, 1-23)
4. Cordgrass Marsh (elevation ~97.5 m) (marsh, natural/artificial, potentially permanent; Figs. 1-24, 1-25)
5. Cattail Marsh (elevation ~97.5 m) (marsh, artificial, potentially permanent; Fig. 1-26)

Private Wetlands

6. South Branch of Rutledge Creek (elevation 91.5–94.5 m) (swamp, natural, permanent, peatland [in part]; Figs. 1-27–1-30)
7. Rutledge Swamp (elevation 94.5–97.5 m) (peat swamp, natural, permanent; Figs. 1-31–1-35)
8. South Soefje Swamp (elevation 94.5–97.5 m) (peat swamp, natural, permanent; Figs. 1-36, 1-37)
9. North Soefje Swamp (elevation 94.5–97.5 m) (peat swamp, natural, permanent; Fig. 1-38)
10. North Soefje Marsh (elevation 96.0–97.5 m) (peat marsh, natural, permanent; Figs. 1-39–1-41)
11. Soefje Cattail Pond (elevation 99 m) (pond, artificial, potentially permanent; Fig. 1-42)

Summary

Five of these wetlands are public, and six are private. There are six swamps, three marshes, one oxbow lake, and one pond. At least five of them, four swamps and one marsh, may be classified as peatlands. Six wetlands are entirely natural in their sources of water, but the two cattail marshes are artificial in both origin and maintenance. The three remaining wetlands exist through combinations of nature and human intervention via wells and/or pipelines installed or maintained by state or private agencies.

Key elevations above sea level:

Red Hill Overlook: >134 m
Ottine village 106.6–109.7 m
Hershop "bog" 103 m
Bogusch's site 97.5–99.0 m
San Marcos River at Ottine (not in flood): ~88.8 m

Overview of the Ottine Wetlands

1. Palmetto Trail Area (Figs. 1-5–1-18)

The namesake species of Palmetto State Park, the dwarf palmetto (*Sabal minor*), may be seen here in abundance (Fig. 1-5), bordering a luxurious, winding path that skirts lagoons (Fig. 1-6) and green ash trees (Fig. 1-7). Part of this wetland is supplied by an artesian well (Fig. 1-8). The water has a neutral pH of 7 and a temperature of 21.6°C and is delivered by the agency of an old ram pump (Fig. 1-9) and by nearly negligible dripping runoff from a wood and stone tower (Figs. 1-8, 1-10). Most of the Palmetto Trail area must rely upon seasonal rains and the flooding San Marcos River to fill the numerous "lagoons" that are its main features. In a survey of Texas state parks this area was described as a "sometime swamp" (Banks, Hollister, and Llewellin 2004). The brief invitation to visitors features a fine color photograph of the native iris (Fig. 1-15) and emphasizes the primordial ambience of the entire region.

In summer the vertical pipe of the artesian well is likely to cease supplying its pulsating fountain of water. Yet some flow usually continues through the ground-level horizontal pump that makes water more reliable here than elsewhere.

The Palmetto Trail area, as small as it is, offers more to the casual visitor per unit time both day and night than any other area accessible to the public. Trees growing here include green ash (*Fraxinus pennsylvanica*), boxelder (*Acer negundo*), hackberry (*Celtis laevigata*), bur oak (*Quercus macrocarpa*), Osage-orange (*Maclura pomifera*), cedar elm (*Ulmus crassifolia*), winged elm (*U. alata*), Shumard oak (*Q. shumardii*), and anaqua (*Ehretia anacua*). Notable shrubs in addition to the dwarf palmetto include buttonbush (*Cephalanthus occidentalis*), red buckeye (*Aesculus pavia*), and rough-leaf dogwood (*Cornus drummondii*).

At the time of this writing one solitary palm tree grew across from the Palmetto Trail in Palmetto State Park (Fig. 1-11). This we believe is the Texas palmetto (*Sabal mexicana*), probably introduced, although the species is thought to be native to the region (Lockett 2003). Its fruits and flowers were not available, so we fell back on microscopic leaf structure as a means to identify the individual to species level (Zona 1990). According to that method of identification (Figs. 1-12, 1-13), the plant is an exotic Mexican palm rather than a native dwarf palmetto or the southeastern cabbage palmetto (*S. palmetto*). Its origin is unknown.

Fig. 1-5
Dwarf palmetto (*Sabal minor*).

Fig. 1-6
Lagoon in the Palmetto Trail area (wetland 1).

Fig. 1-7
Green ash (*Fraxinus pennsylvanica*) strangled by enormous Alabama supplejack vine (*Berchemia scandens*); dwarf palmetto (*Sabal minor*) at lower left (wetland 1).

Fig. 1-8
Artesian well (right foreground) and water storage tower (wetland 1).

Fig. 1-9
Hydraulic ram pump watering lagoon (wetland 1).

Fig. 1-10
Stone roundhouse with statement of palmetto's changing geographic distribution (wetland 1).

A remarkable forb growing along the trails is Florida lettuce (*Lactuca floridana*). In spring and summer it grows to heights exceeding 3.5 m. By this time the ubiquitous swamp katydid (*Amblycorypha oblongifolia*) has molted from juvenile to adult, and a hundred or more individuals of both sexes crowd onto a single plant, where they defoliate the leafy giant until little more than the central stem remains.

Fig. 1-11
Texas palmetto (*Sabal mexicana*) as identified by leaf structure (wetland 1).

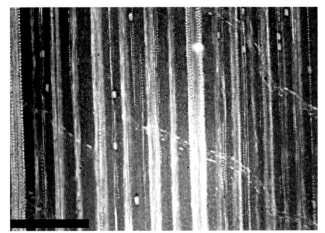

Fig. 1-12
Microscopic leaf structure of dwarf palmetto (*Sabal minor*) (scale bar = 750 μm).

Fig. 1-13
Microscopic leaf structure of Texas palmetto (*Sabal mexicana*) (scale bar = 750 μm).

Fig. 1-14
Yellow flag (*Iris pseudacorus*) (wetland 1).

Fig. 1-15
Dixie iris (*Iris hexagona* var. *flexicaulis*) (wetland 1).

Fig. 1-16
Black vultures (*Coragyps atratus*) coming to roost in winter (wetland 1).

Fig. 1-17
Nine-banded
armadillo (*Dasypus
novemcinctus*) foraging
at night (wetland 1).

Fig. 1-18
Lagoon drying out
(wetland 1).

The most popular botanical attractions in the public wetland are two iris species. Both may be seen with good timing and good luck on or near the Palmetto Trail. The introduced yellow-flowered species known as yellow flag (*Iris pseudacorus*) (Fig. 1-14) is a more reliable sight that seems to be pushing aside the smaller native purple-flowered species known as Dixie iris (*I. hexagona* var. *flexicaulis*) (Fig. 1-15). We found no records of the exotic species in the literature, but there are less-threatened populations of the native growing in swamps on private lands nearby. Occasionally someone confuses the abundant giant spiderwort (*Tradescantia gigantea*) with the less prolific native iris because both bloom early in spring and both bloom in one shade of blue or another.

Other plants along the Palmetto Trail include the red-fruited Carolina wolfberry (*Lycium carolinianum*), white-flowered frostweed (*Verbesina virginica*), poison-ivy (*Toxicodendron radicans*), Virginia-creeper (*Parthenocissus quinquefolia*), and Alabama supplejack (*Berchemia scandens*).

Beginning in late afternoon a racket of flapping wings may be heard in the vicinity of the water tower (Fig. 1-16). These, according to our experience, are black (*Coragyps atratus*) and turkey vultures (*Cathartes aura*) congregating to roost for the night. Hisses, croaks, and the sounds of wing blows fill the night air as the big birds squabble over perching sites high up in the trees. Anyone who walks the trail after dark should be prepared for sounds that are disturbing even when one is sure of their source. Even louder are the crashing noises on the ground made by the nine-banded armadillo (*Dasypus novemcinctus*) as it forages for insects after dark (Fig. 1-17). This immigrant from Mexico seems oblivious to quiet observers and will sometimes walk right up to a pair of boots.

During our studies all of the lagoons dried up during the summer (Fig. 1-18), although the oxbow lake of wetland 3 did not. Precisely the opposite condition was noted by Raun (1958), who reported that the lagoons never went dry even when the oxbow lake was empty. He attributed their longevity to the overflow from a "pumphouse" and water from a "spring." As far as we can tell, this must refer to the water storage tower and to the artesian well, respectively, neither of which was up to the job of keeping even one lagoon wet in the summer of 2001. Raun also seemed to report introduced water-hyacinths that were not present when we arrived. There might be some confusion because he identified them by the scientific name of one of the two iris species that grow among the lagoons.

2. Hiking and River Trail Area (Figs. 1-19–1-21)

With the exception of a few local seeps feeding small ponds of very limited extent, such as that of the extinct mud boil (Fig. 1-19), there is little reliable water here, although there is a creek that has cut a small, canyonlike gorge where a visitor leaving the trail could take a dangerous tumble (Fig. 1-20). The wetland is often nearly dry, at least on the surface.

Closest to the Palmetto Trail is the Hiking Trail; farther beyond is the River Trail. Many of the wetland trees and shrubs may be seen while walking the loops, but some species, including black willow (*Salix nigra*) (Fig. 1-21), bald-cypress (*Taxodium distichum*), Texas persimmon (*Diospyros texana*), cottonwood (*Populus*

Fig. 1-19
Former site of mud boils, now a pond colonized by grass-leaf arrowhead (*Sagittaria graminea*) (wetland 2).

Fig. 1-20
Creek cutting deeply through the soil on its way to
the San Marcos River in the Hiking and River Trail
area (wetland 2).

Fig. 1-21
Black willow (*Salix nigra*) with shelf fungi.

deltoides var. *deltoides*), pecan (*Carya illinoinensis*), mulberry (*Morus rubra*), Osage-orange, sassafras (*Sassafras albidum*), wax-myrtle (*Myrica cerifera*), Roosevelt weed (*Baccharis neglecta*), and elderberry (*Sambucus canadensis*), are more conspicuous elsewhere in the park and on private land.

3. Oxbow Lake Area (Figs. 1-22, 1-23)

This small, shallow lake was formed naturally long ago when the meandering San Marcos River changed course, cutting off and stranding one of its bends (Fig. 1-22). In fact, its name is merely a general term used to designate all such relict riverbeds. Left high and dry by the San Marcos, it relies for its water upon rain, temporary reunion with the river via floods, and two artificial sources. One of these is an artesian well (Fig. 1-23), and at the time of writing the other was a pipeline originating on the property of the Warm Springs Foundation (Park Superintendent Mark Abolafia-Rosenzweig, pers. comm. 2003). The pH of Oxbow Lake is 7.0, or neutral, and the pH of the nearby artesian well is slightly higher at 7.1. The lake is said to be no more than 1.8 m deep in most places, and though it appeared to be stable during our study, it has gone dry in past decades for up to three years at a time (Raun 1958).

Along its shores grow the tall southern reed (*Phragmites australis*), southwestern bristle grass (*Setaria scheelei*), and bald-cypress.

Fig. 1-22
The oxbow lake, viewed from its west end (wetland 3).

Fig. 1-23
Artesian well near oxbow lake (wetland 3).

4. Cordgrass Marsh (Figs. 1-24, 1-25)

In 2001 the Gulf cordgrass marsh (Fig. 1-24) could be supplied at will with water by a pipeline that stretched across the wetland from one side to the other on its way to the oxbow lake, which can be supplemented by the same pipe. The source on the adjacent Warm Springs Foundation property presumably delivered water with a temperature of 21.7°C, though an ancient geothermal well still bubbled when we visited the grounds (Fig. 1-25). Its temperature was 36.7°C and thus only slightly lower than human body temperature. Rains and the San Marcos River in flood are additional sources of water for the cordgrass marsh. The pH of the water here is unique for its alkalinity. Far from the acid condition expected of "boggy" areas, it was slightly higher than 8.0, by far the highest pH we discovered in these wetlands.

This somewhat isolated marsh has been estimated as 4 hectares in extent (Raun 1959). It is located in a shallow basin just southeast of park headquarters as they stood at the time of writing and is easily missed from the road. The dominant

Fig. 1-24
Gulf cordgrass marsh
with Malaise trap in
place (wetland 4).

Fig. 1-25
Old but still bubbling
geothermal well near
cordgrass marsh
(wetland 4).

plant is Gulf cordgrass (*Spartina spartinae*), also known as "sacahuiste," which grows in characteristic clumps in shallow and at least slightly saline water. Here, too, is the nonnative, weedy giant cattail (*Typha domingensis*), which seems to have largely if not entirely replaced the native broad-leaf species (*T. latifolia*) recorded by all those who studied the area before us (Bogusch 1930; Raun 1958; Parks 1935a). Growing nearby is a stand of southern reed, a tall true grass that grows abundantly along the margin of the adjacent oxbow lake. It extends into the cordgrass patch but is uncommon elsewhere. Other plants include Carolina wolfberry, Roosevelt weed, mesquite (*Prosopis glandulosa*), downy hawthorn (*Crataegus mollis*), and switchgrass (*Panicum virgatum*). Most of these may be seen as colonizers or invaders of the marsh.

In the mid–twentieth century this tiny wetland dried out so severely that the clay soil cracked (Raun 1958). We don't know the fate of a second cordgrass meadow that Bogusch studied in the early part of the century. It was located on private land near the northern entrance to Palmetto State Park. Perhaps it has vanished completely.

5. Cattail Marsh (Fig. 1-26)

After familiarizing ourselves with maps and the results of previous studies, we were surprised to find a cattail marsh in the hinterland of Palmetto State Park (Fig. 1-26). Perhaps it was recently formed. There is certainly no mention of this wetland in a previous survey (Raun 1958). It relies for its water on runoff from an artesian well that gushes even in summer on nearby private land. This is the same well that supplies the cattail pond on that property, and we believe that these plants founded the marsh through their progeny. The pH here is 7.1.

Despite the force of the well's flow, the water evaporates and sinks into the ground during the hottest months before it reaches this clearing. The soil then becomes cracked and dry, and thus, at least at the surface, the water in the small artificial wetland is seen to be ephemeral or temporary. It is unclear if the marsh would survive if the well on private property should stop flowing. Perhaps both

Fig. 1-26
Cattail marsh with
Malaise trap in place
(wetland 5).

its maintenance and its origin can be traced to human disturbance that can be viewed, as it almost never is, in a favorable light.

The small patch is less than half a hectare in extent. It is not on any trail and can be reached only by a short hike into the woods. It is dominated by the exotic giant cattail, which was confirmed by our observation of flowering spikes during the wet summer of 2004. Thus the native broad-leaf cattail, the only species recorded previously (Bogusch 1928; Raun 1958), would appear to be extirpated from the Ottine Wetlands because we did not find it at any of the sites where cattails now occur.

In this hinterland cattails flourish alongside sallow caric sedge (*Carex lurida*), which is green in winter when the former plants have died back aboveground to dry brown leaves and stems. Here also is the halberd-leaf hibiscus (*Hibiscus laevis*). Together the low-growing species survive in a clearing surrounded by towering black willows, boxelder, and green ash.

6. South Branch of Rutledge Creek (Figs. 1-27–1-30)

This, the first of the private wetlands treated in detail here, is a complex array of permanent ponds, seeps, shaded swampy patches, a pecan-rich floodplain that receives water from the San Marcos River (Fig. 1-27), and Rutledge Creek that empties into the same (Fig. 1-28). Like all but one of the private lands, its waters arise from natural sources only. We measured a pH of 6.0 in Rutledge Creek, so it can be described as acidic (Fig. 1-29). In such peatlands as these one sometimes experiences the phenomenon known as "quaking." Quaking occurs when the ground beneath the feet quivers like Jell-O with each step and is caused by a thick mat of vegetation floating atop underlying water or saturated, mucky peat.

Much of the land endures periodic heavy flooding from the San Marcos River, and as a result pecan trees flourish in tall stands as they do nowhere else. Trees and shrubs are much like those of the public trails across the highway except that

Fig. 1-27
Floodplain pecan grove (wetland 6).

Fig. 1-28
At the center of the Ottine Wetlands; fork of Rutledge Creek where its north and south branches merge on the way to the San Marcos River (wetland 6).

Fig. 1-29
Rutledge Creek with grape lianas (wetland 6).

Fig. 1-30
Feral hog skull (wetland 6).

black willow, cottonwood, elderberry, and, of course, pecan are more abundant. Feral hogs (*Sus scrofa* Linnaeus), now common in the area (Taylor 1991), moved into the area sometime between the conclusion of the Bogusch and Raun studies and the beginning of our own (Fig. 1-30).

7. Rutledge Swamp (Figs. 1-31–1-35)

Rutledge Swamp is the best example in the Ottine area of a permanent swampland more typical of the southeastern United States. It is supplied by seeps from underlying groundwater in the Carrizo Sands Formation of Eocene age (Fig. 1-31). These yielded a pH of 5.5, one of the lowest or most acidic values measured in the Ottine area, and thus quite opposite in nature from the alkaline waters of the cordgrass marsh. Water temperature was typically and predictably 21.7°C. At the time of our study the muck was treacherously deep, and the going was difficult. Water is liable to enter the tops of sinking knee-high boots with unpleasant consequences, and the extraction of rapidly flooding footwear may prove difficult. The foot itself might tear loose from a boot that remains mired in muck. It is a good idea to travel in pairs in such a wetland.

Fig. 1-31
Seep flowing through
Rutledge Swamp
(wetland 7).

Fig. 1-32
Wind-thrown
Shumard oak (*Quercus
shumardii*) (wetland 7).

Fig. 1-33
Canebrake rattlesnake
(*Crotalus horridus*)
digesting recent meal
beneath fallen willow
log (*Salix nigra*).

Fig. 1-34
Juvenile canebrake
rattlesnake (*Crotalus
horridus*) resting several
feet aboveground.

Fig. 1-35
Juvenile western cottonmouth (*Agkistrodon piscivorus leucostoma*) preparing to strike.

Within the surrounding forested lowland that lies between pasture and swamp, spectacular wind-thrown trees marked the approach to our goal. Among them were giant cottonwoods and Shumard oaks (Fig. 1-32). These fell not because of high winds per se but because of unusually weak root systems that did not penetrate deeply enough into the oxygen-poor soil to provide solid anchors. Not so susceptible to windthrow are the ashes, boxelder, willow, cottonwood, and pecan. Dwarf palmettos are abundant here, as are wax-myrtle and the occasional scythe-fruit arrowhead (*Sagittaria lancifolia*).

Deadly water-hemlock (*Cicuta maculata*) is so poorly anchored in the muck that a boot splashing down near the plant may cause it to lean precariously. If an unsuspecting human should happen to chew on its leaves, stems, or roots, the experience may prove fatal. Dangerous animals that may be encountered here include canebrake rattlesnakes (*Crotalus horridus*) (Figs. 1-33, 1-34) and western cottonmouth (*Agkistrodon piscivorus leucostoma*) (Fig. 1-35).

In a few clearings huge cinnamon ferns (*Osmunda cinnamomea*) grow to heights of almost 2 m in the company of wax-myrtle and yaupon that flourish on drier ground nearby. One expects a dinosaur to peer through such foliage. Cattails were once reported in Rutledge Swamp (Raun 1958), but we saw none in our own explorations. Perhaps these early colonizers have been replaced by other plants.

During the mid-twentieth-century drought Rutledge Swamp dried up (Raun 1958). In conversation with one of its current owners who was aware of that history, we learned that it had not done so during his own lifetime of several decades.

8. South Soefje Swamp (Figs. 1-36, 1-37)

South Soefje Swamp is supplied with seeps from the same rock formation that supplies the more northern Soefje wetlands as well as Rutledge Swamp. Green ash and boxelder grow abundantly, and in a small, open, marshy area, balloon vine (*Cardiospermum halicacabum*) and cocklebur (*Xanthium strumarium*) dominate the scene (Fig. 1-36). Peat has accumulated over thousands of years to depths in

Fig. 1-36
Photographing
cocklebur (*Xanthium
strumarium*) in clearing
(wetland 8).

Fig. 1-37
Sassafras grove
(*Sassafras albidum*)
on upland border
(wetland 8).

some areas exceeding 4.6 m (Graham 1958). Along a slope that is the interface between upland pasture and the swamp we were surprised to find a small sassafras grove that may well lie at the western limit of this species' range (Fig. 1-37).

9. North Soefje Swamp (Fig. 1-38)

Like South Soefje Swamp, this wetland receives a permanent or nearly permanent supply of cool water from numerous seeps in the underlying sand. We found the water more acidic than elsewhere, with a pH as low as 5.0. It flowed from seeps at temperatures ranging from 18.9°C to 22.2°C. Some of this variation might be due to shaded versus more exposed conditions, because a few seeps lie at the base of a wax-myrtle or some other shrub or tree. On the western edge of North Soefje Swamp water accumulates to form ponded, grass-crowded avenues that suggest a Louisiana bayou. Green ash is the defining swampland tree here (Fig. 1-38). It flourishes alongside black willows and boxelder.

Those forested peatlands described in the past as "bogs" lie in this area and in the nearby South Soefje Swamp. In reality there are no bogs anywhere in the Ottine Wetlands because none of these peatlands rely upon precipitation as their sole source of water. Quite to the contrary, in the present case hillside seeps seem to constantly moisten exposed clearings of black peat surrounded by trees and shrubs. Long before our studies began, consideration was given to the possibility of mining these sites commercially (Chelf 1941; Plummer 1941, 1945).

According to Raun (1958, 1959) this swamp and the adjacent North Soefje Marsh dried up during the devastating seven-year drought of the mid–twentieth century. Despite presumably recurrent catastrophes these wetlands have survived

Fig. 1-38
Ash swamp with green ash (*Fraxinus pennsylvanica*) (wetland 9).

much as they are over the long term for at least the last eight thousand years, according to the evidence of pollen samples extracted from peat cores (Graham 1958; Bryant 1977).

10. North Soefje Marsh (Figs. 1-39–1-41)

In this wetland the same seeps that supply the adjacent tree-dominated swamp maintain an open marsh of primarily herbaceous vegetation (Fig. 1-39). More conspicuous here than elsewhere is giant cutgrass (*Zizaniopsis miliacea*) that must often be mistaken at a distance and even at close range for cattail (Figs. 1-40, 1-41).

Fig. 1-39
Cool, acidic seep from Carrizo Sands Formation with overhanging vines of Alabama supplejack (*Berchemia scandens*) (wetland 10).

Fig. 1-40
Working through giant cutgrass (*Zizaniopsis miliacea*) in winter (wetland 10).

Fig. 1-41
Giant cutgrass
marsh with
Drummond rattlebox
shrubs (*Sesbania
drummondii*)
(wetland 10).

The latter does not grow at this site as far as we could tell. If the visitor jumps up and down on thick sedge near the seeps, the quaking phenomenon may be experienced. This is caused by thickly matted vegetation growing atop saturated accumulations of peaty muck.

Alongside giant cutgrass grows sallow caric sedge, southern wax-myrtle, dwarf palmetto, coffee senna (*Senna occidentalis*), halberd-leaf hibiscus, and balloon vine. This wetland was the largest marsh we saw in the Ottine area and one of only a very few of any size.

11. Soefje Cattail Pond (Fig. 1-42)

The cattail pond (Fig. 1-42) is an artificial wetland maintained at the time of writing by a gushing artesian well. Water seeping through the banks of the pond also sustains the nearby publicly owned cattail marsh of Palmetto State Park. Long ago a fish hatchery stood on the spot and was hailed as the most reliable refuge for aquatic life in the midst of droughts that dried up nearly every other source of water (Raun 1958). This insurance is no longer extant.

Fig. 1-42
Giant cattail pond
(*Typha domingensis*)
(wetland 11).

2

Trees and Their Epiphytes and Parasites

The arboreal component of the Ottine Wetlands consists of three distinct associations. These include a wet floodplain with its associated wetlands and a mesic to moist element growing on sandy to rocky slopes abutting the floodplain proper. The upland element of the flora is largely complementary to the two associations presented here and is treated at length in the authors' *Plants of the Texas Lost Pines* (forthcoming). The present work treats the lowland component from mesic slopes to the margins of perennial watercourses, including the various wetlands in between and the aquatic habitat beyond.

Characteristic trees of mesic slopes are Shumard oak (*Quercus shumardii*), sassafras (*Sassafras albidum*), and Mexican plum (*Prunus mexicana*). Floodplain species include boxelder (*Acer negundo*), pecan (*Carya illinoinensis*), smooth hackberry (*Celtis laevigata*), downy hawthorn (*Crataegus mollis*), Texas persimmon (*Diospyros texana*), anaqua (*Ehretia anaqua*), green ash (*Fraxinus pennsylvanica*), black walnut (*Juglans nigra*), Osage-orange (*Maclura pomifera*), chinaberry (*Melia azedarach,* introduced), red mulberry (*Morus rubra*), sycamore (*Platanus occidentalis*), eastern cottonwood (*Populus deltoides*), bur oak (*Q. macrocarpa*), bastard oak (*Q. sinuata* var. *sinuata*), black willow (*Salix nigra*), western soapberry (*Sapindus saponaria* var. *drummondii*), bald-cypress (*Taxodium distichum*), winged elm (*Ulmus alata*), American elm (*U. americana*), and cedar elm (*U. crassifolia*).

Lower ground, such as swamps, marshes, pond margins, peatlands, and seepage slopes, contains combinations of these species depending upon history and chance colonization events. Other species are known to have occurred here in the past but have not been reported recently. Examples from the distant past are mesic sweetgum (*Liquidambar styraciflua*), the floodplain smooth alder (*Alnus serrulata*), and river birch (*Betula nigra*). All three are native to present-day East Texas but are known from Central Texas wetlands only as ancient pollen preserved in peat deposits (Patty 1968).

No less intriguing are tree species reported in the early twentieth century (Bogusch 1928; Parks 1935a) but not reported since. These have since been extirpated or, like sassafras, are extremely rare today. These species include the mesic mockernut hickory (*Carya alba*), parsley hawthorn (*Crataegus marshallii*), pasture hawthorn (*C. spathulata*), the Texas-endemic Sutherland's hawthorn (*C. sutherlandensis*), southern red oak (*Quercus falcata*), Carolina basswood (*Tilia americana* var. *caroliniana*), the floodplain-endemic Texas hawthorn (*C. texana*), and green hawthorn (*C. viridis*).

Most of the plants in these lists are of eastern and southeastern affinity. Notable exceptions include Texas persimmon (western two-thirds of Texas and adjacent Mexico), anaqua (South Texas and Mexico), bur oak (eastern Great Plains, Tallgrass Prairie), western soapberry (south-central United States and Mexico), Sutherland's hawthorn (southern Texas), and Texas hawthorn (coastal Texas). The above are species of more southern or central affinity. Lindheimer's hackberry (*Celtis lindheimeri;* south-central Texas and New Mexico), Mexican ash (*Fraxinus berlandieriana;* south-central Texas and New Mexico), and Texas red oak (*Quercus buckleyi;* Central and west-central Texas) have been reported by others, but we have not seen them. The latter are species of arid arroyos, streams, and canyons of the southern Edwards Plateau, South Texas, and New Mexico. The Texas palmetto (*Sabal mexicana*) occurs in Palmetto State Park, where it appears to be represented by a single specimen.

Epiphytes of these trees include the vascular Spanish-moss (*Tillandsia usneoides*), largely restricted to trees of the floodplain, and the xeric, nearly ubiquitous ball-moss (*T. recurvata*). The hemiparasitic mistletoe (*Phoradendron tomentosum* (*P. serotinum* var. *t.*)) afflicts hackberries and elms of the area in particular.

GYMNOSPERMS (CONIFERS)

Eastern red-cedar
Juniperus virginiana C. Linnaeus
CUPRESSACEAE (CYPRESS FAMILY)
Fig. 2-1
Pollen shed (late February) March–May from male plants bearing tiny yellow cones

Field recognition: Shrubby to arborescent (treelike), evergreen conifer with resinous-aromatic foliage bearing appressed, scalelike leaves. Juvenile foliage needlelike and prickly. Bark reddish-brown, dividing and shredding into thin, fibrous strips or scales with frayed ends. Female plants bearing berrylike, glaucous-green seed cones that are sweet and edible when mature, though resinous in flavor. These ripen blue, with a powdery bloom, during winter.
Similar species: None.

Fig. 2-1
Eastern red-cedar (*Juniperus virginiana*) with berrylike cones.

Remarks: Known to Central Texans as "red-cedar," this conifer, so common in surrounding uplands, is quite rare within the wetlands. It is better suited for invading prairies and old fields where fires are not allowed to burn. One conveniently located specimen was growing alongside the Palmetto Trail in the state park at the time of this writing.

Distribution: Eastern inland half of the United States, East and Southeast Texas to the West Cross Timbers and Balcones Escarpment. It is at the southwestern margin of its range here.

Bald-cypress
Taxodium distichum (C. Linnaeus) L. C. Richard
TAXODIACEAE (REDWOOD FAMILY)
Fig. 2-2
Pollen shed in spring from drooping catkins of tiny male cones

Field recognition: Large deciduous conifer of swamps and watercourses with light green, featherlike shoots bearing flat, needlelike leaves that turn reddish-brown (cinnamon) before falling during autumn. Trunk massive, tapering upward and often swollen, fluted, or buttressed at base with reddish to gray-brown bark, shallowly furrowed into thin fibrous ridges. Seed cone small, globose, green, and resinous, with hard, tight scales.

Similar species: None.

Fig. 2-2
Bald-cypress
(*Taxodium distichum*),
autumn aspect.

Remarks: This and the rarely seen eastern red-cedar (*Juniperus virginiana*) are the only conifers occurring naturally in the wetlands according to our experience. Bald-cypress is an integral part of southeastern swamps, where it may tower 40 m overhead and impress onlookers with a trunk 2.5 m wide. In the Ottine area we saw only a few. These were growing at the edge of a swamp and along the shore of the oxbow lake. All were modest in size. "Knees" rise aboveground from the root system, but their function is debated. Perhaps they obtain oxygen in waterlogged soil, perhaps they provide support, or perhaps it is a combination of both. Bald-cypress received its common name because it sheds its featherlike foliage in the fall, unlike more familiar evergreen conifers.

Distribution: Southeastern U.S. Coastal Plain and Mississippi River valley west to the southern Edwards Plateau of Texas. Frequently cultivated and widely naturalized along watercourses.

ANGIOSPERMS (FLOWERING PLANTS)

Boxelder
Acer negundo C. Linnaeus
ACERACEAE (MAPLE FAMILY)
Fig. 2-3
Flowering February–April

Field recognition: Deciduous tree of stream banks and bottomland woods with opposite, pale green feather-compound leaves. Leaves with 3–7 opposite, egg-shaped, coarsely toothed, or lobed leaflets, including a terminal one. Fruits are characteristic maple "keys"; paired samaras with curved wings swept forward. Bark light brown, smooth to slightly roughened, green on very young twigs and branches.

Similar species: The combination of opposite, pinnate-compound leaves and the winged fruits is diagnostic.

Fig. 2-3
Boxelder (*Acer negundo*) leaves and winged fruits.

Remarks: Boxelder is the only maple tree in this wetland. It reaches 22 m in height and is notable for the manner in which long, smooth, green adventitious shoots sprout directly from the main trunk. Its leaves are sometimes confused with those of poison-ivy, but they are arranged in pairs opposite one another on the stem rather than alternating singly as in poison-ivy.

Distribution: From the eastern inland United States to the Rocky Mountains in the north and as far west as the central United States in the south.

Anaqua

Ehretia anacua (M. Teran & J. Berlandier) I. M. Johnston
BORAGINACEAE (FORGET-ME-NOT FAMILY)
Fig. 2-4
Flowering late February–May; fruiting June (rarely to October)

Field recognition: Small to moderate-sized evergreen to semi-deciduous tree with egg-shaped to elliptic dark green leaves that are harshly sandpapery above. Flowering in late winter and spring, in showy white clusters of flowers with yellow throats. Fruiting in early summer (June), yellow to dark orange berrylike drupes, sweet and edible. A tasty boraginaceous snack. Bark light orange-brown, corky and furrowed into ridges broken into small flat-topped, short-rectangular blocks, similar to green ash (*Fraxinus pennsylvanica*).

Similar species: None.

Remarks: The bark of anaqua is similar to that of ash, but the pattern in the case of mature trees is more blocky and more suggestive of an alligator's hide. Heights of 15 m may be attained. It is the only known host plant for the leaf-eating Texas tortoise beetle (*Coptocycla texana*), though we saw this Texas-endemic insect only on rusty blackhaw (*Viburnum rufidulum*). Anaqua is more of an upland species than many of the trees featured here.

Fig. 2-4
Anaqua (*Ehretia anacua*) flowers.

Distribution: An unusually restricted range: within the United States, only in South Texas north to Hays and Travis counties; elsewhere in the northeastern states of Mexico.

Texas persimmon
Diospyros texana G. Scheele
EBENACEAE (EBONY FAMILY)
Fig. 2-5
Flowering February–June

Fig. 2-5
Texas persimmon (*Diospyros texana*).

Field recognition: Shrub or small deciduous tree with small, nearly stalkless, reverse egg-shaped to oblong, leathery leaves that are hairy below and rolled under along the margins. Flowers white, urn-shaped, unisexual with staminate (male) and pistillate (female) occurring on different plants (dioecious). Ripe berrylike, edible fruits small, black, and finely hairy with a sweet, juicy, blackish-brown pulp. Bark thin and smooth, light gray, peeling in thin layers.

Similar species: Bark and trunk resemble the ornamental crapemyrtle (*Lagerstroemia indica* C. Linnaeus), which has larger, hairless leaves. It also resembles the semi-evergreen farkleberry (*Vaccinium arboreum* H. Marshall), which likewise possesses smooth, glaucous-green foliage quite unlike that of Texas persimmon, which is hairy below. Though reported by Parks (1935a), farkleberry is a sandy upland species at the western margin of its southeastern U.S. range here.

Remarks: Texas persimmon is easily recognized by its very smooth, gray, flaky bark. It may be a tree as tall as 12 m or a much smaller shrub, and its black fruits are enjoyed by wildlife and humans alike.

Distribution: Confined to Texas and northern Mexico. By comparison with its usual upland habitat of open rocky woods, savannas and brushlands, and open slopes and along streams and arroyos throughout the western two-thirds of Texas and adjacent northern Mexico, Texas persimmon seems quite out of place on the rich, alluvial terraces above the San Marcos River at Palmetto State Park. But there it seems to thrive, attaining the size of a small understory tree alongside other, mostly eastern, floodplain species.

Honey mesquite
Prosopis glandulosa J. Torrey
FABACEAE (BEAN FAMILY); MIMOSOIDEAE (MIMOSA SUBFAMILY)
Fig. 2-6
Flowering late April–May(–July or later, rains permitting)

Field recognition: Small to moderate-sized spreading, deep-rooted deciduous tree with paired-compound, drooping, feathery or fernlike foliage that casts a thin, light shade. Flowers borne in short, fingerlike, drooping yellowish-white or greenish catkins with conspicuous exserted stamens. Sweetly fragrant and nectariferous. Fruits long pods becoming pale tan with purple-red streaks or spots when ripe. Seeds surrounded by a sweet-acidic pulp. Bark dark gray to black and developing deep longitudinal furrows and ridges with age. Originally on stream banks and rocky slopes of arid and desert regions, it has spread widely north and eastward into prairies, grasslands, and pastures, where it is often regarded as a pest.

Similar species: None.

Remarks: Most Texans have little difficulty identifying the spiny mesquite tree even when it is bare and without its narrow leaves and catkin flowers that attract a variety of pollinating insects. It reaches heights of 10 m but often grows as a shrub. Mesquite is currently expanding northward into disturbed areas such as abandoned fields and is widely viewed as an unwelcome, invasive alien in habitats like the Ottine Wetlands.

Distribution: Currently throughout the southwestern United States, northern Mexico, and Texas, though rare in extreme East Texas and Louisiana.

Fig. 2-6
Honey mesquite
(*Prosopis glandulosa*)
foliage.

Oaks

Bur oak *Quercus macrocarpa* A. Michaux
Bastard oak *Q. sinuata* T. Walter var. *sinuata*
Shumard oak *Q. shumardii* S. Buckley
FAGACEAE (BEECH FAMILY)
Figs. 1-32 [p. 29], 2-7–2-10
Flowering spring

Field recognition: Bur oak: Large deciduous tree of bottomland forests and prairies in calcareous clays and on limestone. Leaf blades large, with rounded lobes and more or less fiddle-shaped with the deepest sinuses near the middle, nearly reaching the midrib. Lower surfaces woolly. Bark thick, dark gray, and longitudinally furrowed into flat-topped or scaly ridges, and fire resistant. Acorns huge, in a scaly cup with fringed margins.

Similar species: White oak (*Q. alba* C. Linnaeus) was reported by Bogusch (1928) but has not been reported since. Anyone fortunate enough to find one here may recognize it by the more evenly lobed leaves (deep or shallow) that are mostly glabrous below rather than woolly and the much smaller acorns whose cups lack the "bur" or "mossy" fringe of curled scales.

Field recognition: Bastard oak: Deciduous tree of moist bottomlands and riparian areas, with reverse egg-shaped (obovate) to oblong, shallow-wavy to irregularly lobed or toothed leaves. Lower leaf surfaces soft pubescent, woolly. Bark light gray or brown with broad, thin, exfoliating flakes, scales, or plates. Acorns small with shallow bowl- or saucer-shaped cups.

Similar species: Water oak (*Q. nigra* C. Linnaeus), reported recently for the first time from the Ottine Wetlands (Williams and Watson 1978), has smaller, glabrous leaves similarly toothed or lobed, each, however, ending in a tiny hair-like awn or bristle representing the excurrent continuation of a leaf vein. This is a characteristic of the red or black oak group (section Lobatae) to which the

Fig. 2-7
The distinctive leaves of bur oak (*Quercus macrocarpa*).

Fig. 2-8
The giant acorn of bur oak (*Quercus macrocarpa*).

Fig. 2-9
Ancient bastard oak (*Quercus sinuata* var. *sinuata*).

water oak belongs and is lacking in the white oak group (section Quercus) of which the bastard oak is a member. In addition, the bark of the former is dark gray to black, hard, and smooth to shallowly furrowed. This represents the southern margin of this species' range.

Field recognition: Shumard oak: Deciduous tree of mesic slopes to poorly drained bottomlands with deeply lobed leaves, the tips of the lobes bearing several branching, diverging veins excurrent as hairlike awns or bristles. Leaf sinuses closing, with margins converging distally, C-shaped in outline. Tall with straight bole having hard, plated or furrowed dark gray to black bark.

Similar species: Southern red oak (*Q. falcata* A. Michaux) was reported by Bogusch (1928) as well as recently by Williams and Watson (1978). It may be distinguished from Shumard oak by its leaves, which have simple, acutely tipped lobes and "open" sinuses with margins diverging distally, V-shaped in outline, and tapering rather than bluntly rounded or truncate leaf bases as in Shumard oak.

Remarks: The three oaks featured here are the most abundant species according to our experience, though three more are known from the wetlands and their surroundings (Appendix 5). If acorns are available, those of bur oak are immediately recognizable by their large size and by the cup, which bears a fringed margin (Fig. 2-8). The rest of the tree is equally impressive, for at heights of up to 50 m this white oak is one of the tallest in the wetlands.

Shumard oak is a red or black oak that, like the bur oak, prefers moist habitats. Wind-thrown individuals may be seen lying prostrate in the deeper woods along Rutledge Creek. Bastard oak, also known as "Durand's oak," is the least familiar of the three and is more difficult to identify. In fact, one ragged, ancient landmark along the Hiking Trail of Palmetto State Park (Fig. 2-9) was identified in a trail guide as Lacey oak (*Q. laceyi* J. K. Small). However, Lacey oak does not occur here (Nixon and Muller 1992).

Fig. 2-10
Shumard oak (*Quercus shumardii*).

Distribution: All three species are eastern trees that reach their southwestern limits in or near the Ottine Wetlands. Bur oak ranges from the northeastern United States, central Great Plains, and adjacent Canada south to Southeast and south-central Texas. Bastard oak occurs throughout the eastern United States through the eastern third of Texas to the western edge of the Blackland Prairie. Shumard oak is distributed throughout the southeastern U.S. Coastal Plain to the Balcones Escarpment.

Pecan
Carya illinoinensis
(F. von Wangenheim) K. Koch
JUGLANDACEAE (WALNUT FAMILY)
Fig. 2-11
Flowering April

Field recognition: Large deciduous tree with long, alternate, feather-compound leaves bearing 6–16 opposite, toothed, asymmetrical, curved lateral leaflets and a straight terminal leaflet. Distal portion of leaf with larger leaflets than lower, proximal portion. Trunk massive, often enlarged at the base, with light gray or brown bark, ridged with appressed or exfoliating platelike scales. Fruit an elongate ellipsoid, shelled nut enclosed in a 3–4-parted husk that spreads open at maturity (dehiscent).

Similar species: Mockernut hickory (*C. alba* (C. Linnaeus) T. Nuttall *ex* S. Elliot) has been reported from the area (Bogusch 1928; Turner et al. 2003a). It may be distinguished from pecan by its shorter compound leaves with

fewer leaflets (5–7 versus 7–17) that
are straight rather than curved,
and velvety beneath. Its fruits are
smaller and more globose than
those of pecan with very thick
shells and husks. An eastern species, it reaches its southwestern
limit here with some disjunction.

Remarks: Pecan is the state tree of
Texas; in the Ottine Wetlands it
grows most abundantly in groves
on the floodplains of private land.
Mature individuals may exceed 50
m in height with trunks over 2 m
in diameter. Its deeply furrowed
bark might be confused with that
of black willow, but the edible
fruits and compound leaves with
broader, dark green leaflets easily
distinguish the pecan. The related
black walnut (*Juglans nigra* C. Linnaeus) also has long, odd-pinnate-
compound leaves. However, the
distal leaflets, rather than being

Fig. 2-11
Pecan (*Carya illinoinensis*).

larger, are smaller than the median ones and are not curved. Also, the pith
of the twigs and branchlets is chambered rather than solid, and the fruits
are surrounded by a continuous, rindlike husk that does not split open upon
maturity.

Distribution: Upper Mississippi and Ohio River valleys, south and west along
their tributaries through East and Central Texas to the Pecos River.

Sassafras
Sassafras albidum (T. Nuttall) C. Nees von Esenbeck
var. *molle* (C. Rafinesque-Schmaltz) M. Fernald
LAURACEAE (LAUREL FAMILY)
Figs. 1-37 [p. 31], 2-12
Flowering March–April

Field recognition: Small deciduous, dioecious tree with egg-shaped to elliptic leaves of four different shapes; entire, mitten-shaped (left or right), and
trilobed. All parts of the plant spicy-aromatic (like root beer) when crushed.
Variety *molle* has finely hairy leaf undersurfaces. Twigs green, mucilaginous,
spicy-aromatic. Bark red-brown (cinnamon colored), furrowed into small,
shallow ridges. Fruit a blue-black drupe when mature and enclosed at the base
by a cuplike reddish stalk (the pedicel and cupule).

Fig. 2-12
Sassafras (*Sassafras albidum*) leaves.

Similar species: None.

Remarks: Sassafras is presently rare in the wetlands. We discovered a single small grove on a hillside just above the South Soefje Swamp. These trees were small, though the plant is capable of reaching 30 m in height. When its leaves are crushed, an aroma suggestive of root beer is released into the air. However, the tree is no longer used to manufacture that beverage because it contains a compound (safrole) deemed carcinogenic by the FDA.

Distribution: Throughout the eastern United States south and west through Iowa, Arkansas, Missouri, Kansas, and the eastern third of Texas to a disjunct population in the wetlands.

Chinaberry
Melia azedarach C. Linnaeus
MELIACEAE (MAHOGANY FAMILY)
Fig. 2-13
Flowering (March–)April–May

Field recognition: Small deciduous tree, umbrella-shaped or with a bushy, rounded crown of large, dark green, frondlike, twice feather-compound leaves. Flowers in large terminal, open-branching inflorescences, petals lavender with a dark purple stamen tube. Fruit a yellow drupe with opaque, whitish-yellow pulp enclosing a bony stone. Bark dark reddish-brown, splitting into shallow, bifurcating, flat-topped ridges revealing lighter orange inner bark in furrows.

Similar species: Fruits of western soapberry (*Sapindus saponaria* var. *drummondii*) are also yellow or amber stone fruits (drupes) in open, branching clusters, but differ in having translucent rather than opaque pulp surrounding the seed.

Remarks: Chinaberry is an exotic, alien tree that is not (yet) particularly common in the wetlands. It reaches heights of 14 m and has large, readily recog-

Fig. 2-13
Chinaberry (*Melia azedarach*) flowers.

nized frondlike, compound leaves, white flowers, and clusters of fleshy yellow fruits that have been known to kill humans when eaten in small quantities.
Distribution: This Asian adventive has escaped cultivation in the southeastern United States from the Atlantic Ocean to Texas.

Osage-orange
Maclura pomifera
(C. Rafinesque-Schmaltz) C. Schneider
MORACEAE (MULBERRY FAMILY)
Fig. 2-14
Flowering April–May

Field recognition: Medium-sized deciduous, dioecious (male and female flowers on separate plants) tree with spiny twigs and shiny, bright green, egg-shaped to elliptic or lanceolate leaves. Flowers in ball-like heads, white, those on female trees developing into softball-sized, lime-green balls (syncarps) with a convoluted, pebbled surface. Seeds embedded near center in sweetish-tasting pulp with milky, sticky juice. Trunk with shallow to deeply furrowed, orange-brown bark with flat-topped ridges, often exfoliating with age. The underlying wood is bright orange or yellow.
Similar species: None.

Fig. 2-14
Osage-orange (*Maclura pomifera*) with fruit in foreground.

Remarks: Osage-orange is an uncommon tree in the Ottine Wetlands. It may reach heights exceeding 20 m and is especially interesting for its remarkable fruit, its protective thorns, and a history that might soon end without human intervention to compensate for interference of a different kind in the remote past. It is thought that the seeds were dispersed by giant mammals with digestive tracts ample enough to acquire and process the huge edible fruits. When these creatures died out after the ice age, possibly extirpated by people, the tree no longer possessed reliable means of dispersing its seeds (Barlow 2000).

Distribution: During recent presettlement times Osage-orange was restricted to Arkansas, Oklahoma, and the Red River valley of Northeast Texas. It is now scattered widely throughout the eastern and south-central United States and the eastern half of Texas via cultivation and the subsequent passive dispersal of the large fruits.

Red mulberry
Morus rubra C. Linnaeus
MORACEAE (MULBERRY FAMILY)
Fig. 2-15
Flowering (March–)April(–May), fruiting late spring to early summer

Field recognition: Small to medium (large) deciduous tree of stream bottoms with small to large heart-shaped to bi- or trilobed leaves with hairy undersurfaces and rough or smooth upper surfaces. Staminate (male) and pistillate (female) flowers borne in pendant greenish spikes on separate plants. Pistillate spikes, following fertilization, develop into multiple fruits, turning from green through red to deep purple-black when ripe; sweet, edible. Resemble elongate blackberries in appearance. Bark gray-brown, smooth when young breaking into broad, flat plates separated by shallow furrows when older.

Fig. 2-15
Red mulberry (*Morus rubra*).

Similar species: Parks (1935a) reported Carolina basswood (*Tilia americana* C. Linnaeus var. *caroliniana* (P. Miller) L. G. Castiglione, synonym of *T. floridana* J. K. Small) from the Ottine Wetlands. This is a large, mesic forest tree of the eastern United States that reaches its southwestern limit along forested streams and lowlands of the southern Edwards Plateau of Central Texas. Its leaves are superficially similar to those of red mulberry but may be distinguished by their bases, which are asymmetrical and angled with respect to the leafstalk (oblique), rather than symmetrically heart-shaped to truncate (straight across), and which are never lobed as those of mulberry often are.

Remarks: This tree bears broad, rounded leaves that allow instant recognition, and though we never saw an individual so tall, they may reach a height of over 20 m. The flowers are white catkins that mature to edible purple fruits often ruined by the feeding of stinkbugs.

Distribution: From the eastern half of the United States to the eastern Rolling Plains and Edwards Plateau of Texas.

Green ash
Fraxinus pennsylvanica H. Marshall
OLEACEAE (OLIVE FAMILY)
Figs. 1-7 [p. 16], 1-38 [p. 32]
Flowering February–April

Field recognition: Deciduous tree of floodplains and swamps with opposite, feather-compound leaves. Leaves with 5–9 opposite, elliptic to oblong leaflets, including a terminal one (odd-pinnate). Male (staminate) and female (pistillate) flowers borne on separate plants (dioecious). Female plants bear drooping clusters of elongate, single-seeded fruits called samaras, each with a long, spoon-shaped wing. Bark pale brown with furrows broken into light tan, rounded, rectangular corky blocks.

Similar species: Mexican ash (*F. berlandieriana* A. P. de Candolle) was reported from the area by Parks (1935a). It is a species of streams and canyons of the southern Edwards Plateau, South and West Texas, New Mexico, and adjacent Mexico. In addition to habitat it may be distinguished from green ash by its smaller stature, leaves with 3–5 narrow leaflets and the shorter samaras (2.5–3.5 cm versus 3.0–7.5 cm), with the wing enclosing the seed to near its base rather than only to the middle, and by the slightly compressed rather than round seed in cross section.

Remarks: Green ash is the flagship lowland tree of the Ottine swamps just as the loblolly pine (*Pinus taeda* C. Linnaeus) is the defining upland tree of the nearby Lost Pines forest. Green ash grows to heights of 20 m or more and requires the moist soil of floodplains, where it may often be seen growing in the heart of the swamp itself, the base of the tree submerged in water, and with buttresses suggestive of the cypress knees that one expects to find in such a habitat. Other names include "red ash," "water ash," "river ash," and "swamp ash."

Distribution: Green ash is an eastern U.S. species ranging from the Atlantic Ocean to a western limit at the Rocky Mountains. The Ottine Wetlands are at or near the southwestern margin of its range within Texas.

American sycamore
Platanus occidentalis C. Linnaeus
PLATANACEAE (SYCAMORE FAMILY)
Fig. 2-16
Flowering March–April(–May)

Field recognition: Large deciduous
tree of stream bottoms with large,
light green, hairy, shallowly 3–5-
lobed leaves. Bases of leafstalks
completely sheathing axillary buds.
Flowers in globose heads, dangling
on pendulous peduncles (stalks),
developing into hanging, ball-like
seed heads ("button-balls"). Massive trunk with bark exfoliating
in thin plates, creating a mosaic
of smooth chalky white, buff, and
greenish patches, becoming dark
brown, thicker, and fissured into
rounded rectangular plates, scales,
or blocks with age.

Fig. 2-16
American sycamore (*Platanus occidentalis*).

Similar species: None.
Remarks: Sycamores are not common
in these wetlands. They are capable
of growing to 55 m in height and are easily recognized by the flaking outer
bark and the smooth lighter layer underneath. The palmately lobed leaves
bear a resemblance to those of maples.
Distribution: Despite a scientific name that suggests a western distribution,
American sycamore occurs from the eastern United States south and west
through East and Southeast Texas to the Grand Prairie and southern Edwards
Plateau.

Downy hawthorn
Crataegus mollis G. Scheele
ROSACEAE (ROSE FAMILY)
Fig. 2-17
Flowering April; fruiting August–October

Field recognition: Small deciduous tree with relatively large (7.5–10.0 cm),
broad, egg-shaped to rounded-triangular, coarsely toothed leaves that are
densely white-woolly (velvety), especially when young. Flowers with five white
petals and twenty faint yellow, pollen-bearing anthers. Bark of trunk gray-
brown, deeply fissured. With or without long, stout thorns scattered along
the twigs and branches. Small, bright red applelike fruits (pomes) ripen in late
summer and fall.

Fig. 2-17
Downy hawthorn
(*Crataegus mollis*).

Similar species: Three other easily recognized hawthorns have been reported for the Ottine Wetlands. These are all at the southwestern limit or margin of their ranges here.

Parsley hawthorn (*C. marshallii* W. Eggleston) is easily recognized by its parsleylike foliage consisting of broadly egg-shaped to blunt-triangular leaves that are deeply cut or dissected half or more of the way to the midrib. It ranges throughout the southeastern Coastal Plain and Mississippi Embayment through East Texas to a limit near the Ottine Wetlands.

Pasture hawthorn (*C. spathulata* A. Michaux) (Fig. 2-18) has small, narrow, reverse egg-shaped leaves (1 cm by 2 cm) that broaden gradually from a narrow, tapered base and short petiole to a broad rounded or coarsely toothed tip. Its twigs and branches are usually armed with long, stout thorns, and its gray bark is smooth or thinly scaly. The southwestern limit of pasture hawthorn's range is in or near the Ottine Wetlands.

Fig. 2-18
Pasture hawthorn
(*Crataegus spathulata*).

Fig. 2-19
Texas hawthorn
(*Crataegus texana*)
flowers.

Green hawthorn (*C. viridis* C. Linnaeus) is a small floodplain tree with short, sparse thorns and pale gray or mottled, smooth to thin-scaly, flaking bark. Its leaves are moderate-sized (2.5–5.0 cm by 1.3–3.0 cm or more), thin, smooth, elliptic to oblong-egg-shaped, and coarsely toothed to shallowly lobed and widest at or beyond the middle. It is at the southwestern margin of its range here.

Two other species are Texas endemics. Texas hawthorn (*C. texana* S. Buckley) (Fig. 2-19) is a small tree with small, thick, broadly egg-shaped leaves, and flowers with reddish, pollen-bearing anthers. It is endemic in Southeast and coastal Texas. Sutherland's hawthorn (*C. sutherlandensis* C. Sargent) is a thicket-forming shrub or small tree to 4.5 m in height, also with small, ovate leaves and flowers with anthers faintly tinged with pink. It is an endemic denizen of low, rich woods in South Texas.

Remarks: Downy hawthorn may reach a height of 12 m with a trunk diameter exceeding 27 cm. We usually encountered much smaller specimens. Hawthorns are notoriously difficult to identify, and according to one apocryphal story, an expert on the genus, when presented with different branches from the same tree, identified them as different species.

Distribution: Downy hawthorn occurs from the eastern, inland half of the United States to the West Cross Timbers and Edwards Plateau of Texas.

Eastern cottonwood
Populus deltoides W. Barton *ex* H. Marshall subsp. *deltoides*
SALICACEAE (WILLOW FAMILY)
Fig. 2-20
Flowering March–April

Field recognition: Large deciduous tree of practically all watercourses with thick, deltoid (triangular) leaves and coarse, rounded teeth, drooping on laterally flattened stalks that flap noisily in the breeze. Flowers unisexual and borne in staminate and pistillate catkins. Bark dark gray, deeply furrowed into rough, coarse, gray ridges.

Similar species: None.

Remarks: Closely related to black willow and just as majestic is the eastern cottonwood or "alamo." Its greatest size is attained in deep alluvial soils. When these are waterlogged, it is subject to windthrow resulting from its shallow, though widespreading, root system. Eastern cottonwood may exceed 30 m in height, and its massive trunks are fixtures along Rutledge Creek, where wind-thrown individuals sprawl across the floodplain floor. The heavy leaves on lax, flattened petioles make a distinctive flapping sound in the breeze that is reminiscent of the sound of running water.

Distribution: From the eastern United States west to the central and western Great Plains and through the eastern half of Texas.

Fig. 2-20
Eastern cottonwood (*Populus deltoides* var. *deltoides*).

Black willow
Salix nigra H. Marshall
SALICACEAE (WILLOW FAMILY)
Fig. 1-21 [p. 22]
Flowering March–April

Field recognition: Deciduous tree of rivers, streams, lakes, ponds, and seepage areas with drooping, light green foliage composed of narrow, elongate, alternate leaves. Several trunks are often present with dark brown bark that is ruggedly furrowed into smooth, gray, platy ridges.

Similar species: None.

Remarks: This is a massive tree that may exceed 20 m in height. Its bark is deeply channeled, and several trunks may rise from the ground forming a circlet or crown. Willows require more water than most of the trees treated here and are usually encountered in remote hinterlands of the swamp.

Distribution: From the eastern half of the United States through the eastern two-thirds of Texas.

Western soapberry
Sapindus saponaria C. Linnaeus
var. *drummondii* (W. Hooker
& G. Arnott) L. Benson
SAPINDACEAE (SOAPBERRY FAMILY)
Fig. 2-21
Flowering May–June

Field recognition: Shrubby, small deciduous tree with feather-compound leaves composed of 10–19 opposite, smooth-margined, curved, lance- or sickle-shaped leaflets, usually lacking (even-pinnate) or with an offset, reduced terminal leaflet. Flowers white, in open, branching inflorescences. Fruits translucent yellow to amber stone fruits (drupes). Bark rough, gray to tan.

Similar species: The pinnate-compound leaves lacking a terminal leaflet and the toothless, sickle-shaped leaflets will separate soapberry from apparently similar

Fig. 2-21
Western soapberry (*Sapindus saponaria* var. *drummondii*).

plants with odd-pinnate-compound foliage, possessing a well-developed terminal leaflet, such as winged sumac (*Rhus copallinum* C. Linnaeus) and smooth sumac (*R. glabra* C. Linnaeus), reported by Parks (1935a) and Williams and Watson (1978), respectively.

Remarks: Western soapberry is a small, uncommon tree that produces open clusters of poisonous yellow fruit. They contain toxic saponins, chemicals with soaplike properties that have been used as such and as fish poison. It is easy to confuse these clusters with those of chinaberry, but the slender, sickle-shaped leaflets of soapberry distinguish it from that exotic species.

Distribution: Occurring sporadically throughout Texas and the south-central United States.

Gum bumelia, woolly-buckthorn
Sideroxylon (Bumelia) lanuginosum A. Michaux
SAPOTACEAE (CHICLE [CHEWING GUM] FAMILY)
Fig. 2-22
Flowering May–July

Field recognition: Deciduous shrub or small tree of uplands to bottomlands with spine-tipped short-shoots (thorns) and reverse lance-shaped to elliptic

leaves clustered on spur shoots or
alternate. Leaves dark green above,
leathery and white-woolly to tawny
beneath. Flowers yellow-white in
axillary clusters. Fruits purplish-
black, fleshy drupes. Bark smooth
when young, dark red-brown with
dotted lenticels, becoming dark
gray and finely furrowed with age.

Similar species: Virginia live oak
(*Quercus virginiana* P. Miller)
foliage may appear similar but lacks
the pubescent lower surfaces or any
trace of thorns.

Remarks: Gum bumelia might be
mistaken for an oak because of its
leaves and even its bark, but the
superficial similarity stops with the
yellowish-white flowers of the former and the plain greenish catkins
of the latter. It sometimes exceeds
15 m in height but may grow as a
smaller shrub as well. We found
a large, isolated individual at the

Fig. 2-22
Gum bumelia (*Sideroxylon lanuginosum*).

interface between pasture and wetland that attracted a remarkable array of
insects to its leaves, branches, and flowers. It is also host for the spectacular,
metallic-green bumelia borer longhorn beetle (*Plinthocoelium suaveolens*).

Distribution: Through the southeastern United States from the Atlantic Ocean
to Arizona and adjacent Mexico, in bottomlands as well as uplands.

Southern hackberry, sugarberry
Celtis laevigata C. von Willdenow
ULMACEAE (ELM FAMILY)
Fig. 2-23
Flowering March–April

Field recognition: Deciduous tree of rich alluvial bottomlands as well as dry
or rocky slopes with alternate, egg-shaped to lance-shaped, long- or short-
pointed leaves with asymmetrical bases. Mature fruits orange to reddish-
brown stone fruits (drupes) with a thin, sweet, dryish, datelike pulp. Bark
smooth and gray with irregular, erupting corky warts, sometimes connected
into rows or ridges with age. Trees often display evidence of disease, insect
infestation, and parasitism (mistletoe) in the form of witch's brooms, galls,
and abnormal growths or swellings on trunks, branches, twigs, and leaves.

Similar species: Lindheimer's hackberry (*C. lindheimeri* G. Engelmann *ex* K. Koch) was reported
by both Bogusch (1928) and Parks
(1935a) for the region. It is a tree of
ravines and canyons of the southern Edwards Plateau and brushlands south of San Antonio, Texas,
and northern Mexico. Apart from
habitat, it may be distinguished
from both smooth and net-leaf
varieties of hackberry by the softly
woolly undersurfaces of its leaves
(rather than smooth to coarsely
pubescent) and its light brown
rather than reddish to orange
drupes.

Remarks: Mature hackberry may
often be identified by its corky,
warty bark. It reaches heights of
over 30 m, bears small dark orange
to reddish-brown edible fruits,
and along with the black willow,
is extraordinarily susceptible to
insect attack as well as attack

Fig. 2-23
Southern hackberry (*Celtis laevigata*) with poison-
ivy vine (*Toxicodendron radicans*).

by the parasitic mistletoe (*Phoradendron tomentosum*). A large but poorly
known herbivorous insect is the Tamaulipas walkingstick (*Diapheromera
tamaulipensis*) (Fig. 2-24), which may often be seen in mating pairs among

Fig. 2-24
Tamaulipas
walkingstick
(*Diapheromera
tamaulipensis*);
adult female on oak.

leaves and branches. Some botanists split this hackberry species in two by recognizing its varieties as full-fledged species: rough- or net-leaf hackberry (*C. reticulata* J. Torrey) and smooth hackberry (*C. laevigata*).

Distribution: Considered as a single species, the tree ranges from coast to coast across the southern half to two-thirds of the United States, with the net-leaf form occupying the western half of Texas, and the smooth form the eastern two-thirds of the state.

Elms
American elm *Ulmus americana* C. Linnaeus
Cedar elm *U. crassifolia* T. Nuttall
Winged elm *U. alata* A. Michaux
ULMACEAE (ELM FAMILY)
Figs. 2-25–2-27
American elm and winged elm: Flowering February–March
Cedar elm: Flowering late July–October, unique among these elms

Field recognition: Deciduous trees with alternate, short-stalked leaves, doubly saw-toothed with pinnate venation and asymmetrical bases. Fruits are diagnostic, disklike samaras; flattened seeds surrounded by a broadly rounded, apically notched, membranous wing.

Field recognition: American elm: A tall, spreading, vase-shaped bottomland tree with moderate-sized smooth leaves. Spring flowering. Bark dark gray or brown with deep black, elongate furrows and bifurcating ridges.

Field recognition: Cedar elm: Tall tree of drier uplands as well as lowlands with small, stiff, thick, dark green, short- or blunt-pointed leaves with sandpapery

Fig. 2-25
American elm (*Ulmus americana*) leaves.

Fig. 2-26
Cedar elm (*Ulmus crassifolia*) with mistletoe (*Phoradendron tomentosum*; center of photo).

Fig. 2-27
Winged elm (*Ulmus alata*).

upper surfaces. Late summer to fall blooming. Bark light brown to gray with shallow, longitudinal furrows and narrow ridges or flat, scaly plates. Twigs with or without corky wings.

Field recognition: Winged elm, wahoo: Small tree of lowlands or uplands with small, thin, light green, long- or sharp-pointed leaves with smooth upper surfaces. Spring flowering. Bark light to dark brown, shallowly furrowed and with exfoliating, curling flakes or plates.

Similar species: Slippery elm (*U. rubra* G. H. Muhlenberg) has never been reported from the Ottine Wetlands, even though the habitat would seem to be to its liking. It may be distinguished from American elm by its larger leaves with sandpapery upper surfaces and its orange-brown, longitudinally furrowed bark, the furrows showing the diagnostic orange inner bark. A moisture-loving tree of lowlands in the eastern half of the United States, it seems to reach its southwestern limit only several kilometers to the north and east of the Ottine Wetlands.

Remarks: We saw three of the four native Texas elms in the Ottine Wetlands. The fourth species, red elm or slippery elm, was observed in the nearby Lost Pines forest but, oddly enough, not in this habitat. Nor was this tree reported by any botanical studies prior to our own. Cedar elm is more abundant than the others. It grows to a height of 25 m. American elm may grow taller, to a height of 30 m, and its leaves are much larger than those of the other two species. Winged elm is the shortest of the three. It usually does not exceed 15 m in height.

Distribution: Of the three, American elm has the greatest range as well as the greatest height and largest leaves. It occurs throughout the eastern United States and southern Canada west to the Rocky Mountains in the north and in the south through East and Southeast Texas to the West Cross Timbers and Edwards Plateau. Winged elm ranges through the southeastern quarter of the United States west through East and Southeast Texas to the East Cross Timbers and Edwards Plateau. Cedar elm has a peculiar, restricted range: from the Mississippi River valley through the eastern half of Texas with a disjunct population in the Suwanee River valley of northern Florida.

EPIPHYTES

Ball-moss
Tillandsia recurvata (C. Linnaeus) C. Linnaeus
BROMELIACEAE (PINEAPPLE FAMILY)
Fig. 2-28
Flowering year-round

Field recognition: Gray-green epiphytic plant in dense, ball-like clumps on trees, rocks, stone walls, tombstones, and utility wires. Stems short, contracted, and concealed by the overlapping bases of the long leaves. Flowers borne on long stalks.

Similar species: The tiny seedlings on tree bark are said to be sometimes confused with lichens (Diggs, Lipscomb, and O'Kennon 1999).

Fig. 2-28
Ball-moss (*Tillandsia recurvata*).

Remarks: Ball-moss grows more abundantly on trees in these wetlands than does its close relative Spanish-moss, and its common name reflects a globular rather than hanging habitus. It also differs by the possession of roots and the position of its flower at the end of a long stalk.

Distribution: Able to endure arid climes, ball-moss grows from the Atlantic Ocean to Arizona and south into the Tropics and beyond, as far south of the equator as Argentina and Chile. Range increase northward has been reported in recent decades, possibly correlated with slight increases in average rainfall (McWilliams 1992).

Spanish-moss
Tillandsia usneoides (C. Linnaeus)
C. Linnaeus
BROMELIACEAE (PINEAPPLE FAMILY)
Figure 2-29
Flowering February–June

Fig. 2-29
Spanish-moss (*Tillandsia usneoides*).

Field recognition: Slender, pendant, gray-green epiphytic plant with long, threadlike leaves arising along branching stems growing on tree branches or utility wires. Roots absent. Flowers stalkless among the leaves.

Similar species: The only plant with which Spanish-moss might be confused is the southern beard lichen (*Usnea trichodea* Ach.), which we have seen to the north in Bastrop County near the Colorado River and which may occur nearby. Unlike that of the flowering vascular plant, the body (or thallus) of this lichen is not differentiated into stems and leaves.

Remarks: Spanish-moss is a rootless flowering plant that hangs from the branches of a variety of trees, such as the hackberry shown in the photograph. Unlike mistletoe, it is not parasitic. The Ottine Wetlands lie near the western edge of this epiphyte's geographic range, and for that reason it is not as abundant or luxurious as one might expect in a swampy habitat.

Distribution: Spanish-moss ranges along the southeastern U.S. Coastal Plain from Virginia south and west through Southeast and South Texas, southward into the Tropics and beyond to Argentina.

PARASITES

Mistletoe
Phoradendron tomentosum (A. P. de Candolle)
G. Engelmann *ex* A. Gray (*P. serotinum*
(C. Rafinesque-Schmaltz) M. C. Johnston var. *tomentosum*)
VISCACEAE (MISTLETOE FAMILY)
Fig. 2-30
Flowering October–March

Field recognition: Yellow-green, bushy, evergreen, dioecious hemiparasite with leathery leaves and brittle stems emerging from the often swollen portions of its host plants. Especially conspicuous on deciduous trees during winter. Female plants produce white, translucent berries (drupes) containing single seeds surrounded by sticky pulp containing viscin threads.

Similar species: None.

Remarks: Unlike Spanish-moss and ball-moss, this flowering plant is parasitic on its host tree. It contains chlorophyll and is apparently a parasite of the wood (xylem) or water-conducting tissue of the host plant, taking water and inorganic nutrients from its host. It attacks hackberry, ash, elm, cottonwood, mesquite, oak, willow, and other trees growing abundantly in the wetlands and adjacent uplands. The sticky seeds are dispersed by birds via bill-wiping, as well as by defecation via cloacal-wiping and the excretion of necklacelike strings of seeds that whirl like bolas as they fall, increasing their chances of intercepting and wrapping around a twig or branch on their way down. The white fruits are toxic to humans.

Distribution: Mistletoe occurs from California to Arkansas and throughout Texas and is thus one of the relatively few western plants of this area.

Fig. 2-30
Mistletoe
(*Phoradendron
tomentosum*).

3
Shrubs and Vines

The shrub and vine component of the lowland flora mirrors the arboreal component in its habitat distribution. Shrubs characteristic of mesic and seepage slopes abutting the floodplain include eastern baccharis (*Baccharis halimifolia*), American beautyberry (*Callicarpa americana*), coral bean (*Erythrina herbacea*), yaupon (*Ilex vomitoria*), pokeweed (*Phytolacca americana*), Carolina buckthorn (*Rhamnus caroliniana*), and rusty blackhaw (*Viburnum rufidulum*). Mesic vines include swallow-wort (*Cynanchum laeve*), anglepod (*Matelea gonocarpos*), climbing boneset (*Mikania scandens*), southern dewberry (*Rubus trivialis*), and fiddle-leaf greenbrier (*Smilax bona-nox*).

Characteristic floodplain shrubs are red buckeye (*Aesculus pavia*), bastard indigo (*Amorpha fruticosa*), buttonbush (*Cephalanthus occidentalis*), rough-leaf dogwood (*Cornus drummondii*), halberd-leaf hibiscus (*Hibiscus laevis*), possumhaw (*Ilex decidua*), Carolina wolfberry (*Lycium carolinianum*), southern wax-myrtle (*Morella* (*Myrica*) *cerifera*), dwarf palmetto (*Sabal minor*), elderberry (*Sambucus canadensis*), Drummond's rattlebox (*Sesbania drummondii*), and bladderpod (*Sesbania* (*Glottidium*) *vesicaria*).

Floodplain vines feature pepper vine (*Ampelopsis arborea*), raccoon-grape (*A. cordata*), Alabama supplejack (*Berchemia scandens*), trumpet-creeper (*Campsis radicans*), balloon vine (*Cardiospermum halicacabum*), leatherflower (*Clematis pitcheri*), Carolina snailseed (*Cocculus carolinus*), dodder (*Cuscuta* sp.), drooping melonette (*Melothria pendula*), Virginia-creeper (*Parthenocissus quinquefolia*), catbrier (*Smilax glauca*), bullbrier (*S. bona-nox*), poison-ivy (*Toxicodendron radicans*), sweet grape (*Vitis cinerea*), mustang grape (*V. mustangensis*), and fox grape (*V. vulpina*).

Notable shrubs recorded in the past (Bogusch 1928; Parks 1935a) but not reported recently are staggerbush (*Lyonia mariana*), an eastern species of moist sand, bogs, and peatlands that reaches its western limit here; and possumhaw viburnum (*Viburnum nudum*), a mesic species of the southeastern United States and East Texas.

As it is for the trees, the prevailing biogeographic pattern is one of eastern mesic and wetland species at or near the western limit of their ranges with a few West and South Texas Tamaulipan elements making their appearance, especially in the drier, subxeric uplands. Examples of the latter among shrubs are the prairie Roosevelt weed (*Baccharis neglecta*), white honeysuckle (*Lonicera albiflora*), Tamaulipan huisache (*Acacia farnesiana* (*A. minuata*)), Texas snakewood (*Colubrina texensis*), bluewood (*Condalia hookeri*), pencil cactus (*Opuntia leptocaulis*), Texas almond (*Prunus texana*), and lotebush (*Ziziphus obtusifolius*).

Examples from among vines are prairie buffalo gourd (*Cucurbita foetidissima*), the Central Texas-endemic scarlet clematis (*Clematis texensis*), the South Texas-endemic short-crowned milkvine (*Matelea brevicoronata*), the South Texas and Tamaulipan swanflower (*Aristolochia erecta*) and American snout-bean (*Rhynchosia americana*).

SHRUBS

Dwarf palmetto
Sabal minor (N. von Jacquin) C. Persoon
ARECACEAE (PALM FAMILY)
Fig. 3-1
Flowering June

Field recognition: Small evergreen palm with large, fan-shaped leaf blades arising at ground level from a subterranean, rarely emergent trunk. Flowers and fruits in long, exserted panicles. Fruits small, dull black with thin, sweet pulp surrounding large seed (1 cm diameter). A denizen of stream bottoms and swamps.

Similar species: One specimen of Texas palmetto (*S. mexicana* K. von Martius) was found growing along the road in Palmetto State Park, apparently planted as an ornamental. It is distinguished from dwarf palmetto by its tall, emergent trunk (to 15 m in height) and larger, longer, feather- or frondlike (costapalmate) rather than fanlike (palmate) leaves that have an elongate, tapering midrib (hastulum) to 10 cm long versus a short (<5 cm), blunt midrib

Fig. 3-1
Dwarf palmetto (*Sabal minor*) along margin of lagoon with fallen ash near foreground.

in the dwarf species. Microscopic details of the leaf venation (Zona 1990) permitted us to determine that this specimen was not the Florida cabbage palmetto (*S. palmetto* (T. Walter) C. Loddiges *ex* J. A. Schultes & J. H. Schultes) (Figs. 1-12, 1-13 [p. 18]).

Remarks: Dwarf palmetto, also known as "swamp palmetto," is the namesake plant of Palmetto State Park. It is a true palm that seldom develops a trunk aboveground and, unlike the Old World date palm (*Phoenix dactylifera* C. Linnaeus), is rarely found as a tree. Nevertheless, it can develop a 2 m trunk (Corner 1966; Zona 1990; Henderson, Galeano, and Bernal 1995). We are not aware of any such individuals in the Ottine Wetlands. There is a single Texas palmetto, with a large trunk (Fig. 1-11 [p. 18]), that was probably introduced into Palmetto State Park, though the species is now believed to be native to south-central Texas (Lockett 2003).

Dwarf palmetto flowers in spring and summer and produces grapelike fruits in fall. This species appeared in the area with the expansion of eastern plants and animals during a wet era during the ice age. As the region dried out, the flora and fauna retreated, leaving relicts behind in suitable habitats such as those underlain by the Carrizo Sands, which supply the wetland with groundwater.

Distribution: Dwarf palmetto currently ranges from the Atlantic and Gulf Coastal Plains of North Carolina south and west to Southeast and South Texas with a western limit in the central part of the state not far west of Ottine.

Roosevelt weed
Baccharis neglecta N. Britton
ASTERACEAE (DAISY FAMILY); ASTEREAE (ASTER TRIBE)
Fig. 3-2
Flowering September–November

Field recognition: Shrub with narrow, linear, elliptic to reverse lance-shaped, gray-green leaves, smooth-margined to toothed and 1-nerved. Twigs green, stems green or striped, with thin ridges. Flowers small, tubular, and clustered in unisexual flower heads on separate plants. Fruits plume-topped dry seeds (achenes), dispersed by wind or water. Growing in limey, calcareous soils.

Similar species: The salt-tolerant eastern baccharis (*B. halimifolia* C. Linnaeus) (Fig. 3-3) of sandy areas on the Atlantic and Gulf Coastal Plain, inland to Oklahoma, north-central Texas, and reaching its southwestern limit in the Ottine Wetlands differs in having broader, elliptic to reverse egg- or diamond-shaped, ragged or weakly toothed leaves.

Remarks: This late-flowering shrub grows on the borders of the Gulf cordgrass marsh but was seldom noticed elsewhere in the wetlands. It is abundant in old fields and pastures along roadside fence lines as shown in the photograph. Eastern baccharis was found growing in Palmetto State Park along a shaded seepage rivulet between the old fish hatchery pond and the cattail marsh.

Distribution: From the Atlantic Ocean to Arizona and south to Mexico.

Fig. 3-2
Roosevelt weed
(*Baccharis neglecta*).

Fig. 3-3
Eastern baccharis
(*Baccharis halimifolia*).

Hollies

Yaupon *Ilex vomitoria* W. Aiton
Possumhaw *I. decidua* T. Walter
AQUIFOLIACEAE (HOLLY FAMILY)
Figs. 3-4, 3-5
Yaupon: Flowering April
Possumhaw: Flowering March–May

Field recognition: Yaupon: Densely, divaricately branching, evergreen shrub
with small, elliptical, scalloped, dark green, leathery leaves. Short, axillary
(occurring in the angles of stem leaves) shoots sometimes spine-tipped.
Pollen-bearing (male) and fruit-bearing (female) flowers borne on separate
plants (dioecious). Bark of trunk and branches smooth, light greenish-gray,
usually adorned with a patchwork mosaic of crustose, powdery (embedded),
and script lichens. Flowers and berrylike fruits (drupes) densely clustered
along axillary twigs, becoming bright red in fall and winter.

Fig. 3-4
Yaupon (*Ilex vomitoria*) leaves.

Fig. 3-5
Possumhaw (*Ilex decidua*) leaves and berries.

Field recognition: Possumhaw: Lighter green, thinner leaves that are deciduous rather than evergreen and taper gradually near the base (cuneate) to a short leafstalk, whereas leaves of yaupon abruptly narrowed (obtuse), broadly rounded, or truncate at the base. Possumhaw less densely branched, with fewer axillary twigs and short shoots along its main axes on which its flowers and orange to red fruits (drupes) are less densely clustered.

Similar species: None.

Remarks: The two hollies occurring in the wetlands are small trees or shrubs reaching heights of 8–10 m, though in our experience they are typically much smaller. Yaupon is distinguished from possumhaw by the former's evergreen condition and leaves, which do not taper extensively toward the petiole. The bright orange to red berries are poisonous to humans, but a ceremonial emetic made from the leaves by American Indians gave the yaupon its scientific name (Diggs, Lipscomb, and O'Kennon 1999). The dense, divaricate branching pattern seems to be an adaptive response to the browsing of herbivores or pruning by the elements (wind, salt spray, freezing, etc.). Indeed, for this reason yaupon is especially favored for ornamental use in hedges and plant sculptures. The leaves of both species are loaded with caffeine and may be toxic to herbivores in large quantities. In fact, the South American *I. paraguariensis* St. Hillaire (Yerba maté) is the source of the dried leaves that are steeped in hot water to make the caffeinated, mildly stimulating beverage known as *maté*, the national drink of Brazil.

Distribution: Both shrubs range from the southeastern United States west through East to Southeast and Central Texas. American holly (*I. opaca* D. Solander), though not reported from the Ottine Wetlands, has apparently been collected elsewhere in Gonzales County (Turner et al. 2003a). This marks the southwestern limit and an apparent disjunction for this small, evergreen tree of the eastern United States. With its rigid, leathery, dark green, prickly foliage and red "berries" of late fall and winter, it is an unmistakable symbol of Christmastime.

Pencil cactus
Opuntia leptocaulis A. P. de Candolle
CACTACEAE (CACTUS FAMILY)
Fig. 3-6
Flowering April–May

Field recognition: Jointed cactus with narrow, cylindrical segments and one long spine per spine cluster (areole). Flowers small with green, yellow, or bronzy petals. Fruits bright red, juicy ovoid berries, persisting long after maturing. Also known as "Christmas cactus." Usually of heavy clay and alluvial, bottomland soils.

Similar species: None.

Remarks: Parks (1935a) reported this species, and we observed it growing in openings on the high (9 m) floodplain terrace of the San Marcos River in Palmetto State Park. It appears to have a preference for heavy clay and alluvial, bottomland soils. We encountered only a few specimens, one of them growing close to the extinct mud-boil pond in Palmetto State Park, where a cactus might not have been expected.

Fig. 3-6
Pencil cactus (*Opuntia leptocaulis*).

Distribution: Also known as "rat-tail cactus," this species has an unusually western distribution for a plant of the Ottine area, ranging eastward from Arizona to Oklahoma and the western two-thirds of Texas and northern Mexico.

Plains prickly-pear, grassland prickly-pear
Opuntia macrorhiza G. Engelmann
CACTACEAE (CACTUS FAMILY)
Fig. 3-7
Flowering May–June

Field recognition: Low, prostrate, clump-forming perennial, pad-bearing (prickly-pear) cactus from thickened, tuberous main roots with glaucous-blue-green, spiny pads. One to six spines per spine cluster (areole), mostly deflexed downward and on upper areoles. Flowers showy with yellow petals

Fig. 3-7
Plains prickly-pear
(*Opuntia macrorhiza*)
flower with bee-
eater assassin bug
(*Apiomerus spissipes*)
feeding on insect prey.

tinged red or orange basally. Fruits reddish-purple when ripe and known as "prickly-pears." Areoles with many tiny urticating (irritating), barbed, hair-like, easily detached spines (glochidia). In sandy or rocky to loamy and clayey soils of plains and grasslands, often hidden among taller grasses.

Similar species: Eastern prickly-pear (*O. humifusa* (C. Rafinesque-Schmaltz) C. Rafinesque-Schmaltz) is a species more characteristic of dry sandy and rocky soils and outcrops than the former species. It may be distinguished by its green (not glaucous) pads with mostly spineless areoles or with at most one spine each, mostly in the upper portion of the pads. It also has fibrous rather than tuberous roots and flowers that are usually all yellow, without red or orange centers and open as early as April. It ranges throughout the eastern half of the United States and Texas. Texas prickly-pear (*O. engelmannii* J. Salm-Reifferscheid-Dyck var. *lindheimeri* (G. Engelmann) B. Parfitt & D. Pinkava) is a much larger, upright shrub that also occurs here in pastures and along roadsides in sandy and rocky uplands to silty or clayey floodplain soils. Its larger stature and uniformly spiny pads will separate it from the former two species.

Remarks: Plains prickly-pear is not a wetland plant, but it does grow abundantly in the periodically inundated floodplain of the San Marcos River near Rutledge Creek. It is tempting to speculate whether the thickened, tuberous main roots of this species are an adaptation to the finer, droughty soils in which it often dwells. Frequently lurking within the yellow and red cup-shaped flower is the predatory bee-eater assassin bug (*Apiomerus spissipes*), often feeding on its latest victim (Fig. 3-7).

Distribution: Plains prickly-pear ranges through the western two-thirds of the United States from southern Michigan and Louisiana westward throughout Texas to California.

Common elderberry
Sambucus canadensis C. Linnaeus
CAPRIFOLIACEAE (HONEYSUCKLE FAMILY)
Fig. 3-8
Flowering May–June(–September)

Field recognition: Large, shrubby, perennial herb with large, opposite, feather-compound leaves composed of egg-shaped-elliptic to lance-shaped, shiny, finely toothed leaflets. Flowers creamy-white, in large terminal, flat-topped clusters (corymbs), showy. Fruits purple-black, berrylike drupes. Cooked fruits edible, used in jellies and wines. A denizen of swamp edges, stream bottoms, and ditches.

Similar species: Its opposite leaves will serve to distinguish elderberry from other plants with pinnate-compound foliage, like sumacs (*Rhus* spp.), with which it could be confused.

Remarks: Elderberry typically grows as a shrub along streams and along the edges of swamps, but it may achieve tree status at heights of up to 4 m or more. We found it especially common along Rutledge Creek. When cooked, the fruits are edible, though the remainder of the plant is poisonous (Diggs, Lipscomb, and O'Kennon 1999).

Distribution: From the eastern half of the United States and southern Canada through the eastern half of Texas, scattered westward.

Fig. 3-8
Common elderberry
(*Sambucus canadensis*)
in flower.

Rusty blackhaw
Viburnum rufidulum C. Rafinesque-Schmaltz
CAPRIFOLIACEAE (HONEYSUCKLE FAMILY)
Fig. 3-9
Flowering (March–)April(–May)

Field recognition: Shrub to small deciduous tree with shiny, bright green, leathery, elliptic to reverse egg-shaped, finely toothed leaves having scurfy, reddish pubescence on petiole and along midrib. Flowers creamy-white, appearing in April in dense, branched, flat-topped clusters. Fruits powdery, blue-black drupes having sweet, edible pulp with a raisinlike flavor. Known also as "wild-raisin." Bark blackish-brown, checkered.

Similar species: Blackhaw (*V. prunifolium* C. Linnaeus) was reported by Parks (1935a) for the Ottine Wetlands region. Its occurrence here would be a surprising disjunction for this species of the eastern United States north of the Coastal Plain. This report may refer to possumhaw viburnum (*V. nudum* C. Linnaeus), however, a southeastern Coastal Plain species that occurs west through East Texas. It has slightly thinner, dull leaves that lack the rufous tomentum of its close relative, the rusty blackhaw.

Remarks: This plant may grow as a shrub or as a tree up to 13 m in height. Its leathery leaves, shiny on the upper surface, are distinctive. In Palmetto State Park we found the Texas-endemic Texas tortoise beetle (*Coptocycla texana*) on the leaves of rusty blackhaw. Perhaps it is a previously unrecognized host plant, or perhaps the insects were overwintering but not feeding.

Distribution: Rusty blackhaw occurs along tree lines near streams from the southeastern United States through the eastern half of Texas and is close to the southern margin of its range here.

Fig. 3-9
Rusty blackhaw
(*Viburnum rufidulum*)
flowers.

Rough-leaf dogwood
Cornus drummondii C. von Meyer
CORNACEAE (DOGWOOD FAMILY)
Fig. 3-10
Flowering May

Field recognition: Deciduous shrub or small bushy tree with opposite, smooth-margined, simple egg-shaped to elliptic-lance-shaped leaves and leaf veins that arise in pairs on opposite sides of the midrib and curve conspicuously forward toward the tip, paralleling the leaf margin. The large, water-conducting vascular elements (vessels) of these veins contain spiral, cellulose thickenings that unravel as fine whitish threads when the leaf is torn or pulled apart (the so-called *Cornus* test). Upper leaf surfaces rough-sandpapery. Flowers small, creamy-white, 4-petaled in terminal, round-topped clusters (cymes). Berrylike fruits (drupes) white when mature. Twigs gray to dark brown, reddish. Of stream bottoms, pond margins, and thickets.

Similar species: None.

Remarks: This plant of at least periodically wet terrain, also known as "Drummond's dogwood," is usually a shrub. Sometimes it grows as a small tree. Its abundant white flowers are favorite nectaring and pollen-gathering grounds for flies, beetles, and wasps. The white, lipid-rich fruits are relished by migrant and overwintering birds during fall and winter.

Distribution: The southeastern United States from Alabama through the eastern half of Texas.

Fig. 3-10
Rough-leaf dogwood
(*Cornus drummondii*)
flowers with pollen-eating fire-collared
beetle (*Batyle
ignicollis*).

Illinois bundle-flower
Desmanthus illinoensis (A. Michaux)
C. MacMillan *ex* B. Robinson & M. Fernald
FABACEAE (LEGUME OR PEA FAMILY);
MIMOSOIDEAE (MIMOSA SUBFAMILY)
Fig. 3-11
Flowering late May–June(–September)

Field recognition: Perennial herb to subshrub from a woody root crown with erect to spreading unarmed stems to 1.5 m high, and alternate, twice-compound, fine, fernlike foliage. Flowers borne in small, axillary, pufflike heads, each with five creamy-white stamens. Fruits short, broad, flattened, sickle-shaped pods (legumes) containing 4–5 seeds and borne in dense, ball-like heads. Turning dark brownish-black when splitting open and persisting to the next growing season. Usually in clay soils of ditches, stream bottoms, fields, roadsides, and other low moist areas.

Similar species: This is the only bundle-flower (*Desmanthus* sp.) in Texas with short, broad, curved legumes. Other Texas species have heads of a few to several elongate, linear, more or less straight pods.

Remarks: Illinois bundle-flower, along with many others in the Mimosa subfamily, displays what are known as nighttime "sleep" (nyctinastic) movements of its leaves. At night the leaflets fold up tightly against their midribs, and these, in turn, fold along the main rachis, effectively "closing up" the leaf. The adaptive significance of this behavior is unclear, though it also occurs immediately in response to physical disturbances such as touching or browsing in some other species of the subfamily, known as "sensitive plants."

Fig. 3-11
Illinois bundleflower
(*Desmanthus
illinoensis*).

False indigo
Amorpha fruticosa C. Linnaeus
FABACEAE (BEAN FAMILY); PAPILIONOIDEAE
(PEA SUBFAMILY)
Fig. 3-12
Flowering April–May

Field recognition: Shrub with dark
green, feather-compound leaves
having many aromatic, gland-
dotted, oblong to elliptic leaflets.
Flowers small in dense, clustered
spikes, reduced-papilionaceous
with single upper, dark blue to
reddish-purple petal (banner),
contrasting with the bright orange-
yellow pollen-bearing anthers.
Fruit small, short, stubby, curved
pods conspicuously gland-blistered
and aromatic.

Similar species: None.

Fig. 3-12
False indigo (*Amorpha fruticosa*) flowers.

Remarks: We saw purple-flowered
false or bastard indigo along a
tree line separating Rutledge Creek from the adjacent cattle pasture, as well
as along the Hiking Trail near the mud-boil ponds in Palmetto State Park.
It grows to a height of 3 m, is poisonous to livestock, and is not a common
sight in the border between upland and wetland, perhaps because it is better
adapted to limestone-derived soils than sand.

Distribution: Widely scattered throughout much of the United States.

Coral bean
Erythrina herbacea C. Linnaeus
FABACEAE (BEAN FAMILY); PAPILIONOIDEAE (PEA SUBFAMILY)
Fig. 3-13
Flowering April–June

Field recognition: Small shrub to subshrub (with stems dying back to the
ground during most winters) from a woody base and root crown with prickly,
ascending stems and remote, alternate, 3-foliate compound leaves of eared-
trilobed to broadly and bluntly triangular leaflets. Flowers in terminal, spikelike clusters on elongate stalks, very showy, scarlet-red, appearing tubular
(though modified-papilionaceous, with a single, long, inrolled upper petal)
(3.0–5.3 cm long). Fruit a blackish legume pod, constricted between seeds,
late-dehiscent and containing bright scarlet seeds. Of low sandy woods and
floodplains along streams and thickets of the Coastal Plain. Often cultivated.

Similar species: None.

Remarks: We found the beautiful scarlet flowers of coral bean in clearings near Rutledge Swamp and along wooded roadsides near the river. The beans themselves are not edible; they are used as rat poison in Mexico (Diggs, Lipscomb, and O'Kennon 1999). This is another flower that appears ideally adapted for pollination by hummingbirds.

Distribution: Coral bean is a species of the Atlantic and Gulf coasts, occurring from North Carolina to Veracruz, Mexico, and in Texas about as far inland as Gonzales County in the Ottine Wetlands (Correll and Johnston 1970) or slightly to the west (Turner et al. 2003a).

Fig. 3-13
Coral bean (*Erythrina herbacea*).

The Sesbanias
Rattlebush *Sesbania drummondii* (P. Rydberg) V. Cory
Bladderpod *S. vesicaria* (N. von Jacquin) S. Elliot
FABACEAE (BEAN FAMILY); PAPILIONOIDEAE (PEA SUBFAMILY)
Figs. 3-14, 3-15
Rattlebush: Flowering June–September
Bladderpod: Flowering August–September

Field recognition: Rattlebush: Basally woody shrub with long, feathery, compound leaves bearing many small, oblong leaflets. Flowers papilionaceous (similar in form to those of peas and beans), yellow with red-orange lines, in elongate clusters arising from the angles between leaf and stem. Fruits conspicuous 4-winged, short-beaked pods in which the loose seeds rattle when dry, often through the winter. Grows in moist areas.

Similar species: See bladderpod (*S. vesicaria*).

Field recognition: Bladderpod: Large, herbaceous (nonwoody) annual with feathery, compound leaves of small oblong leaflets. Flowers papilionaceous, dark reddish-orange. Fruit an inflated, baglike pod tapering at both ends, containing a few seeds, presumably adapted for aquatic dispersal via flotation.

Similar species: See rattlebush (*S. drummondii*).

Remarks: Unlike its close relative, the green-stemmed annual bladderpod, rattlebush is a perennial woody plant. Its pods also differ by containing more (poisonous) seeds and by rattling if shaken when dry.

Fig. 3-14
Rattlebush (*Sesbania drummondii*) flowers.

Distribution: Rattlebush occurs along the southeastern Coastal Plain from Florida to Veracruz, Mexico. Bladderpod, also known as "bagpod," occurs from the Atlantic Ocean to Texas, tending to grow near the coast, and is also known by the scientific name *Glottidium vesicaria* (N. von Jacquin) R. Harper. It may be adventive in the coastal southeastern United States from the West Indies.

Fig. 3-15
Bladderpod (*Sesbania (Glottidium) vesicaria*).

Red buckeye

Aesculus pavia C. Linnaeus
HIPPOCASTANACEAE (HORSE-CHESTNUT FAMILY)
Fig. 3-16
Flowering March–May

Field recognition: Large deciduous shrub or small tree having large, dark green, opposite, compound leaves with long, lance-shaped-elliptic to reverse lance-shaped, finely toothed leaflets arising from a common point at the leaf base (palmate), like a fan or the spokes of a wheel. Snapdragon-like, tubular red flowers in large, showy, terminal, erect, branched, pagoda-like clusters. Inedible seeds with large, shiny rich chestnut-brown coats and large, pale, round scars in tan, leathery capsules. Bark smooth, gray-brown.

Similar species: Mexican buckeye (*Ungnadia speciosa* S. Endlicher), an upland species of rocky limestone canyons, slopes, and ridges in South, Central, and West Texas and adjacent New Mexico and Mexico, was reported by Parks (1935a). Belonging to a closely related family (Sapindaceae), it shares a similar floral morphology and fruit structure with the true buckeyes (*Aesculus* spp.). However, it has alternate, feather-compound leaves with opposite leaflets, including a terminal leaflet at the tip (odd-pinnate).

Remarks: Red buckeye usually grows as a shrub, but it may reach the status of a small tree. This plant flowers quickly in early spring, and its bright red flowers are favored by hummingbirds and tiger swallowtails (*Papilio glaucus* C. Linnaeus).

Distribution: From the Atlantic Ocean to the Edwards Plateau of Central Texas.

Fig. 3-16
Red buckeye (*Aesculus pavia*) flower.

Fig. 3-17
Halberd-leaf rose-
mallow (*Hibiscus
laevis*).

Halberd-leaf rose-mallow
Hibiscus laevis C. Allioni
MALVACEAE (MALLOW FAMILY)
Fig. 3-17
Flowering May–November

Field recognition: Large, shrubby, hairless, perennial herb with long-pointed, triangular-eared, arrowhead-shaped leaves with short, divergent basal lobes. Flowers large, showy, with a central, pollen-bearing stamen column and five exserted, knobbed stigmas. Five petals, white or pink with crimson-purple blotch at the base. Fruit erect, smooth capsules, splitting open when dry to reveal round seeds with short reddish-brown hairs. Of marshes, shallow water, and other open, wet areas.

Similar species: None.

Remarks: This white-petaled, red-throated relative of okra is a bona fide wetland plant that we encountered in North Soefje Marsh and in the cattail marsh of Palmetto State Park. It is better known by the scientific name *H. militaris* A. Canavilles.

Distribution: Throughout the eastern half of the United States west through East and Southeast Texas to the Rolling Plains and Edwards Plateau.

Southern wax-myrtle
Morella (Myrica) cerifera (C. Linnaeus) J. K. Small
MYRICACEAE (BAYBERRY FAMILY)
Fig. 3-18
Flowering March–April

Field recognition: Evergreen shrub or small tree with narrow, reverse lance-shaped, alternate leaves dotted with aromatic, fine yellow, powdery glands. Flowers unisexual in short, erect, axillary, conelike catkins on separate plants.

Fig. 3-18
Southern wax-myrtle (*Morella* (*Myrica*) *cerifera*) ringing an extinct wetland flat.

Berrylike fruits (drupes) of pistillate (female) plants with a whitish-blue waxy coating containing palmitic acid. Bark smooth, dark gray.

Similar species: None.

Remarks: Wax-myrtle is an evergreen shrub or small tree, rarely reaching heights up to 12 m and preferring boggy ground. It is adapted to nutrient-poor (oligotrophic), acidic wetland conditions and has root nodules that contain nitrogen-fixing microorganisms. It is especially abundant on private wetlands in the company of cinnamon ferns, though usually on banks, hummocks, or drier soil nearby. Its waxy fruits are rich in palmitic acid and are relished as a high-energy food by yellow-rumped (formerly known as myrtle) warblers (*Dendroica coronata*) (Martin, Zim, and Nelson 1961).

Distribution: From the southeastern United States west to East and Southeast Texas, at the western margin of its range here.

Pokeweed
Phytolacca americana C. Linnaeus
PHYTOLACCACEAE (POKEWEED FAMILY)
Fig. 3-19
Flowering Late June–September(–October)

Field recognition: Large, shrubby, perennial herb from a large rootstock with reddish or purplish stems and large, light green, elliptic to egg-shaped or lance-shaped, hairless, stalked leaves. Small white to greenish-white or pink flowers borne in erect spikelike clusters in the angles of the leaves, elongating and becoming lax and nodding as the berries mature to a dark purple color. Grows in rich, low ground, stream bottom woods, thickets, and disturbed areas such as recent clearings and roadsides.

Similar species: None.

Remarks: Pokeweed is a potentially lethal plant that is eaten in some parts of the South after very young leaves are boiled repeatedly. The boiled water itself

Fig. 3-19
Pokeweed (*Phytolacca americana*).

becomes poisonous as a result. In fact, this white-flowered species should not even be handled without gloves, according to some recent sources (Diggs, Lipscomb, and O'Kennon 1999), though we handled it without ill effects.

Distribution: Through the eastern United States and southern Canada west to East, Southeast, and Central Texas.

Buttonbush
Cephalanthus occidentalis C. Linnaeus
RUBIACEAE (COFFEE FAMILY)
Fig. 3-20
Flowering June–July(–September)

Field recognition: Shrub or small tree with opposite or whorled elliptic to lance-shaped, smooth-margined leaves and white, tubular flowers borne in dense, fragrant, globose heads with conspicuously exserted styles. A denizen of swamps and pond or stream margins.

Fig. 3-20
Buttonbush flower
(*Cephalanthus occidentalis*).

Similar species: None.

Remarks: Buttonbush grows as a shrub or small tree less than 6 m in height. Individuals may be seen in and among the lagoons along the Palmetto Trail. The white flowers are favorite nectar sources for swallowtail butterflies. The flowers are arranged in a sphere that resembles a sea urchin because of their many protruding styles.

Distribution: Buttonbush occurs throughout the United States and southern Canada.

Carolina wolfberry
Lycium carolinianum T. Walter
SOLANACEAE (POTATO FAMILY)
Fig. 3-21
Flowering January–November

Field recognition: Thorny, spreading evergreen shrub to about 2 m in height with erect to sprawling branches and succulent (fleshy), grayish-green, glabrous, short-linear to rounded or wedge-shaped leaves in densely whorled clusters about the thorns (modified short-shoots). Flowers borne in the leaf angles, open bell-shaped and lavender to purple with darker purple streaks leading into the throat. Fruits red-orange ovoid berries, pendant from subtending green stalks with leafy calyces,

Fig. 3-21
Carolina wolfberry (*Lycium carolinianum*).

suggesting tiny tomatoes or chili peppers. Of saline coastal wetlands, around ponds, ditches, and marshes as well as in drier sandy or gravelly soils of brushlands and hills in South Texas. Evidently its adaptations to the physiological drought of brackish and saline coastal wetlands have preadapted (or exapted) it to the physical drought of more arid regions. Indeed, several other close relatives (*Lycium* spp.) occur in West Texas and the desert Southwest and adjacent Mexico.

Similar species: None.

Remarks: Carolina wolfberry is a purple-flowered plant that produces numerous and unmistakable bright red berries that cause the plant to bend with their numbers and weight. It grows best near ponds, in saline soils, and in marshlands, precisely the habitats in which we encountered it.

Distribution: From Mississippi south and west through coastal and South Texas to northeastern Mexico. The species is disjunct at its inland limit within the Ottine Wetlands.

American beautyberry
Callicarpa americana C. Linnaeus
VERBENACEAE (VERVAIN FAMILY)
Fig. 3-22
Flowering June–July; fruiting in fall, winter

Field recognition: Bushy, deciduous shrub with scurfy, ill-smelling, aromatic foliage of opposite, egg-shaped to elliptic leaves with toothed or scalloped margins. Flowers in dense clusters (cymes) in the angles of the leaves, pinkish, lavender to white. Fruits globose, berrylike drupes in dense clusters, very showy, maturing in fall to shiny violet or red-purple, occasionally white in var. *lactea* F. von Mueller. A denizen of moist lowland woods, thickets, edges of swamps, and sometimes dryish uplands.

Fig. 3-22
American beautyberry
(*Callicarpa americana*)
flowers.

Similar species: None.

Remarks: Beautyberry is classified as a lowland shrub, but it is also an important feature of nearby dry, sandy forest. We saw a few individuals in the vicinity of Rutledge Swamp, where their presence surprised us in such close proximity to cinnamon ferns and water-hemlock. One of these displayed its pinkish flowers only a few moments' walk from the muck.

Distribution: From the Atlantic Ocean to Texas.

VINES

Fiddle-leaf greenbrier, bullbrier
Smilax bona-nox C. Linnaeus
SMILACACEAE (GREENBRIER FAMILY)
Fig. 3-23
Flowering April–May

Field recognition: Tough, tendril-bearing vine with green, prickly stems. Often forming dense, impenetrable thickets. Leaves tardily deciduous, bright green, often mottled or blotched with pale green to whitish areas, variable in size and shape, ranging from small and triangular with indented sides to large and egg-shaped or heart-shaped with sides curving outward. Leaf edges with fine prickles or bristles and a thickened, veinlike margin. Tendrils arising in pairs from near the bases of the leaf petioles. Flowers greenish-yellow, in stalked, flat-topped clusters arising from the angles between the leafstalks and stem (axils). Berries ripen to a glaucous-blue-black in fall.

Similar species: Greenbrier or wild sarsaparilla (*S. glauca* Walter) was reported from the area by Parks (1935a). It may be distinguished by its leaves, which are strikingly whitish or powdery blue-gray below, contrasting with their green upper surfaces, and their smooth edges with their margins that are not indented as in fiddle-leaf greenbrier. It ranges throughout the eastern United

Fig. 3-23
The leaves and stems of fiddle-leaf greenbrier (*Smilax bona-nox*).

States through East Texas, and has not been reported from these wetlands recently.

Bristly greenbrier (*S. tamnoides* C. Linnaeus) ranges throughout the eastern half of the United States and Texas and has been collected nearby in Gonzales County (Turner et al. 2003b). It may be distinguished from the former greenbriers by its weak, bristlelike, dark rather than pale prickles, and leaves with thin margins that are not indented along the sides and that are not pale-glaucous below.

Remarks: Fiddle-leaf greenbrier is the most widespread and abundant greenbrier in Texas and sooner or later becomes a leg-snagging impediment when one moves through wooded areas of these wetlands. The soft, young shoots of this and other greenbrier species, however, may be eaten raw or cooked as a vegetable, having a pleasant, asparagus-like flavor with a slight bitterness. It is a species of moist to dry woods and thickets.

Distribution: Fiddle-leaf greenbrier ranges through the southeastern United States, west and south through the eastern two-thirds of Texas into Mexico.

Poison-ivy
Toxicodendron radicans (C. Linnaeus) K. E. O. Kuntze
ANACARDIACEAE (SUMAC FAMILY)
Fig. 3-24
Flowering mid-April–May (and later)

Field recognition: High-climbing vine with aerial rootlets (or low to high shrub with creeping, subterranean stems or rhizomes). Leaves alternate and trifoliate with egg-shaped to elliptic or reverse egg-shaped to diamond-shaped, irregularly toothed or lobed leaflets. Small yellow-green flowers borne in erect, branched axillary or lateral clusters. Fruits white to cream-colored drupes ripening in fall. Foliage turns bright red in late summer or early fall. Of lowland woods, floodplains, and streamside thickets. All parts of poison-

Fig. 3-24
The alternately placed leaves of poison-ivy (*Toxicodendron radicans*).

ivy contain urushiols, chemicals causing a delayed contact-dermatitis reaction in sensitive or sensitized individuals.

Similar species: Eastern poison-oak (*T. diversilobum* (J. Torrey & A. Gray) E. Greene), a species of upland, open sandy woods abutting the Ottine Wetlands, differs in having leaflets with broader, blunter teeth and/or shallow, rounded lobes that are pubescent beneath. The fruits are also hairy. It is a low, creeping shrub from horizontal, subterranean stems. Unfortunately, it is similar to poison-ivy in its toxic properties. Boxelder (*Acer negundo*) has opposite rather than alternate leaves.

Remarks: Poison-ivy's caustic, allergenic secretion is a force to be reckoned with in Central Texas, whether the habitat is upland forest or lowland swamp. The plant has yellow-green flowers and may grow as an herbaceous vine, a woody vine, or even as a woody shrub. The leaves resemble those of boxelder, but they alternate singly along the stem as shown in the photograph and are not arranged opposite one another as they are on boxelder's branches.

Distribution: From Nova Scotia to British Columbia south to Florida, Arizona, and Oaxaca, Mexico (Correll and Johnston 1970); throughout Texas.

Climbing boneset
Mikania scandens (C. Linnaeus) C. von Willdenow
ASTERACEAE (DAISY FAMILY); EUPATORIEAE (BONESET TRIBE)
Fig. 3-25
Flowering August–November

Field recognition: Perennial twining vine with opposite, stalked, egg-shaped to triangular, long-tipped, smooth-margined to shallowly angled or lobed, basally heart-shaped leaves. Flower heads borne in irregular, rounded, densely branched, flat-topped clusters at the tips of axillary branches; rays lacking. Flowers whitish or rarely pink-tinged. Fruits 5-ribbed, blackish dry seeds (achenes) crowned with a pappus of fine, stiff bristles. Growing in river bottoms, stream "galleries," lowland moist woods, and seepage slopes.

Fig. 3-25
Climbing boneset
(*Mikania scandens*).

Similar species: None.

Remarks: Climbing boneset, also known as "hempweed," is a vine that climbs over other plants despite its membership in the sunflower family. It bears white and pink flower heads and is described as "infrequent" in Texas (Correll and Johnston 1970). Our experience in the nearby Lost Pines forest corroborates that appraisal, for we saw only one specimen in years of study. It was growing across a creek, as were many of the more abundant individuals we have seen in the Ottine Wetlands.

Distribution: Occurring widely in warmer areas of the Americas. From the northeastern United States south and west through East and Southeast Texas to the East Cross Timbers and Edwards Plateau, and south into Mexico.

Trumpet-creeper
Campsis radicans (C. Linnaeus) B. Seemann *ex* E. Bureau
BIGNONIACEAE (CATALPA FAMILY)
Fig. 3-26
Flowering (May–)June–August(–October)

Field recognition: High-climbing (to over 10 m), deciduous vine with aerial rootlets and opposite, feather-compound leaves of 9–11 egg-shaped to egg-shaped-oblong or elliptic-lance-shaped, toothed leaflets, including a terminal one. Flowers in terminal, open-branched clusters, stout, very showy with a leathery orange calyx and tubular-funnelform orange to orange-red floral tube and orange-red to red limb (mouth) (6–9 cm long). Spreading limb and throat scarlet-orange. Fruits long (9–12 cm) cylindric-oblong, tapering pods, dark brown when mature. Splitting along longitudinal sutures and packed with flattened, brown, 2-winged seeds. Of stream banks, floodplains, lowland woods, and disturbed areas such as fencerows. Commonly cultivated but invasive and aggressive.

Similar species: Pepper vine (*Ampelopsis arborea* (C. Linnaeus) Koehne) is another bushy to high-climbing vine with compound leaves. However, its

Fig. 3-26
Trumpet-creeper
(*Campsis radicans*).

leaves are twice- to thrice-compound and arranged alternately along the stems rather than opposite as in trumpet-creeper. Its small, greenish flowers appear from late June to August, giving rise to black berries with an unpleasant bitter-pungent, peppery flavor in the fall. It ranges throughout the southeastern United States, through the eastern half of Texas to northeastern Mexico.

Remarks: Trumpet-creeper is a vine with beautiful, red, tubular flowers that are pollinated by hummingbirds. Dermatitis reportedly may result from the handling of its leaves and flowers (Diggs, Lipscomb, and O'Kennon 1999).

Distribution: Trumpet-creeper winds around tree trunks and crawls over and through shrubs from throughout the southeastern United States through the eastern half of Texas and scattered as an adventive escaped from cultivation elsewhere.

Drooping melonette, meloncito
Melothria pendula C. Linnaeus
CUCURBITACEAE (MELON FAMILY)
Fig. 3-27
Flowering (March–)May–November

Field recognition: Slender, tendril-bearing herbaceous vine from a perennial root, with 5-angled or lobed leaves and small yellow or greenish unisexual flowers solitary or clustered in the leaf axils, the staminate and pistillate flowers mimicking one another. Pendant unripe fruits resemble tiny watermelons, ripening to a purplish-black color. Climbs over shrubs and small trees at swamp edges and moist, wooded, sandy to loamy soils.

Similar species: None.

Remarks: Meloncito is a climbing vine with yellow flowers that offer only pollen as a reward to potential pollinators, usually bees. Female flowers apparently mimic their pollen-producing male counterparts in order to entice visitation,

Fig. 3-27
Drooping melonette
(*Melothria pendula*).

effecting cross-pollination. The ripe fruits are reputed to be violently emetic in humans.

Distribution: Through the southeastern quarter of the United States, west through East and Southeast Texas to the Rolling Plains and Edwards Plateau, and south to Mexico.

Carolina snailseed
Cocculus carolinus (C. Linnaeus) A. P. de Candolle
MENISPERMACEAE (MOONSEED FAMILY)
Fig. 3-28
Flowering June–July(–October)

Field recognition: Unarmed perennial, tendril-less vine with variable leathery leaves, generally egg-shaped-triangular to variously lobed or nearly round, and minutely rough-puberulent above. Flowers small, greenish-yellow, unisexual, in elongate clusters dangling from the leaf angles. Fruits translucent red drupes containing a ridged, snail-shaped stone. Of thickets, lowland woods, and disturbed areas.

Similar species: Carolina snailseed may be distinguished from greenbriers (*Smilax* spp.) by its lack of tendrils or prickles and the pubescent, finely sandpapery leaves.

Remarks: The photograph shows the bright red fruits of snailseed rather than its yellow or greenish flowers. The fruit is "said to be edible but we haven't tried it" (Correll and Johnston 1970); nor have we. Also known as "Carolina moonseed."

Distribution: This vine occurs throughout the southeastern United States west through the eastern half of Texas.

Fig. 3-28
Fruits of Carolina snailseed (*Cocculus carolinus*).

Leatherflower
Clematis pitcheri J. Torrey & A. Gray
RANUNCULACEAE (BUTTERCUP FAMILY)
Fig. 3-29
Flowering late April–early July(–September)

Fig. 3-29
Leatherflower (*Clematis pitcheri*).

Field recognition: Herbaceous vine with an angled or ribbed stem clambering over shrubs in stream bottom thickets, in open woods, and along fencerows. Leaves opposite and feather-compound with 3–5 egg-shaped, simple to lobed, net-veined leaflets, and terminating in a slender, tendril-like filament (a modified leaflet). Flowers ovoid to urn-shaped and nodding, of four sepals, dull purple or red-purple, ribbed outside and dark purple inside, recurved or spreading at the tips. Fruit a head of long-tailed achenes clothed in flattened, silvery hairs.

Similar species: Texas virgin's bower (*C. drummondii* J. Torrey & A. Gray) is an upland vine of fencerows, canyons, arroyos, pastures, scrubby slopes, and disturbed areas in Central, South, and West Texas and the southwestern United States and northern Mexico. It differs in having smaller, coarsely toothed leaflets, small, whitish, nonleathery sepals that are wheel-like or widespreading, and achenes with erect-hairy, plumose styles ("tails"), the heads of which look like feathery balls. Bogusch (1928) reported scarlet or Texas clematis (*C. texensis* S. Buckley) in these wetlands. It is an endemic upland species of rocky limestone cliffs, slopes, canyons, and stream banks of the southern Edwards Plateau and is unique in the genus for having red, leathery sepals.

Remarks: Leatherflower's nodding flowers bloom on a vine that grows near creeks and along fence lines. The nodding floral habit has resulted in its other common name, "bluebell." The more familiar bluebell is a member of a different family of plants (the Campanulaceae), hence another (and perhaps preferable) name for the species treated here is "leatherflower."

Distribution: Leatherflower's range is unusual compared to those of most plants in the region, for it occurs from the eastern Great Plains from Illinois and Indiana south and west through north-central, Southeast, Central, South, and West Texas and New Mexico to northern Mexico rather than from the Atlantic Coastal Plain to Texas.

Alabama supplejack
Berchemia scandens (J. Hill) K. Koch
RHAMNACEAE (BUCKTHORN FAMILY)
Fig. 3-30
Flowering late April–early May

Field recognition: High-climbing, smooth, unarmed, twining liana (woody
vine) with smooth dark greenish to grayish bark and reddish to yellowish-
brown naked twigs. Leaves alternate, deciduous, egg-shaped to oblong-elliptic
with conspicuous featherlike venation of parallel secondary veins branching from the midrib. Flowers small, yellowish or greenish-white and borne
in small spikelike or branched clusters at the ends of short lateral shoots.
Fruits small, ellipsoid blue-black, olivelike drupes. Of lowland wooded slopes,
stream bottom thickets, and forest edges.

Similar species: None.

Remarks: This is a woody vine with pinnately veined leaves and whitish flowers. We encountered it in wetlands and on upland slopes leading down to the
swamps and marshes where massive individuals seem to strangle the trunks
of green ash trees.

Distribution: The southeastern United States through East, Southeast, and
north-central Texas and in ravines of the southern Edwards Plateau; also
occurring in southern Mexico and Guatemala.

Fig. 3-30
Alabama supplejack
(*Berchemia scandens*).

Southern dewberry
Rubus trivialis A. Michaux (*R. riograndis* L. Bailey)
ROSACEAE (ROSE FAMILY)
Fig. 3-31
Flowering late March–April; fruiting May–early June

Field recognition: Trailing or low-arching, semi-evergreen (first-year shoots or primocanes), woody perennial with recurved-prickly and red glandular-bristly canes and leafstalks. Leaves approximately fan-compound (with stalk of terminal leaflet longer), of five oblong to lance-shaped, coarsely toothed leaflets, smooth except for main veins below. Flowers borne in lateral, erect second-year shoots (floricanes) in few-flowered open clusters (cymes). Leaflets of flowering canes smaller than those of the primocanes and with leaves deciduous. Flowers roselike on bristly-glandular pedicels (flower stalks) with five ephemeral white petals surrounding many central ovaries. Fruit an aggregate of single-seeded drupelets (the familiar raspberry), changing color from green through red to black when ripe. Edible and delicious.

Similar species: None.

Remarks: Dewberries are edible after they change color from green to red to black-purple, but like those of red mulberry they attract stinkbugs that sometimes ruin the flavor. Abundant fruit is the positive side of a trade-off that includes prickly, trailing stems ready to trip up a passing boot.

Distribution: The white-petaled flowers of southern dewberry appear in a variety of habitats throughout the southeastern United States west through East and Southeast Texas to the West Cross Timbers and on to the Edwards Plateau.

Fig. 3-31
Southern dewberry
(*Rubus trivialis*).

Balloon vine
Cardiospermum halicacabum C. Linnaeus
SAPINDACEAE (SOAPBERRY FAMILY)
Fig. 3-32
Flowering June–November

Field recognition: Herbaceous, annual vine usually sprawling over weeds
and bushes or trailing along the ground in streambeds and low, disturbed
areas. Leaves twice-trilobed, the leaflets toothed or lobed and lance-shaped
to narrowly egg-shaped or diamond-shaped, with long-pointed tips. The
small, complex white to yellowish-green flowers are borne in few-flowered,
branched clusters (corymbs) at the ends of slender, tendril-bearing stalks in
the angles (axils) of the leaves. Small flowers develop following successful
fertilization into surprisingly large, membranous, inflated 3-celled pods.
Similar species: None.
Remarks: Balloon vine, named for its remarkably inflated fruit, blooms in
greenish-white flowers and covers the ground in open, marshy areas. It is the
host plant of the soapberry bug (*Jadera haematoloma*).
Distribution: A native of the eastern half of Texas, though some believe it has
been introduced, and in other warm climes of the New World. Balloon vine is
sometimes cultivated as an ornamental and is widespread in warmer regions
of the Western Hemisphere.

Fig. 3-32
The flower of balloon
vine (*Cardiospermum
halicacabum*).

Virginia-creeper
Parthenocissus quinquefolia
(C. Linnaeus) J. Planchon
VITACEAE (GRAPE FAMILY)
Fig. 3-33
Flowering May–July

Fig. 3-33
Virginia-creeper (*Parthenocissus quinquefolia*).

Field recognition: Perennial, deciduous vine, high-climbing via adhesive disks on branching tendrils. Leaves fan-compound of five (sometimes six) oblong-obovate to elliptic, coarsely toothed acuminate leaflets. Flowers many (to two hundred), small and yellow or greenish in branched clusters (cymes). Fruits dark blue or black, glaucous berries. Leaves turning bright red in fall, coinciding with ripening of the fruits. Of diverse habitats. Best development in lowland woods and along streams.

Similar species: Seven-leaf creeper (*P. heptaphylla* (S. Buckley) N. Britton *ex* J. K. Small), a species endemic to the Edwards Plateau and Lampasas Cut Plain of Central Texas, was reported for the area by Parks (1935a). It is an upland or canyon species of rocky or sandy soils clambering over small trees and shrubs and bearing smaller, succulent leaves of mostly seven leaflets.

Remarks: Virginia-creeper is a climbing vine with greenish flowers that may twine so thickly about the trunk of a tree that the bark is nearly hidden from view. It is often confused with poison-ivy, but in this case the problem is not a skin rash. It is potential illness from eating the fruits.

Distribution: Virginia-creeper ranges throughout the eastern United States west to Minnesota and through the eastern half of Texas.

Grapes

Fox, winter grape *Vitis vulpina*
C. Linnaeus
Sweet grape *V. cinerea* (G. Engelmann)
G. Engelmann *ex* P. M. Millardet
Mustang grape *V. mustangensis*
S. Buckley
VITACEAE (GRAPE FAMILY)
Fig. 3-34
*Fox, wintering grape: Flowering late April–
mid-May; fruiting October–November
Sweet grape: Flowering May–early June;
fruiting September–November
Mustang grape: Flowering
late March–April; fruiting late
June–August(–September)*

Fig. 3-34
Sweet grape (*Vitis cinerea*).

Field recognition: Fox grape: High-climbing, perennial vine with a woody trunk. Bark light red-brown, finely splitting into narrow, shallow ridges and often flaking in thin, narrow strips. Leaves broadly heart-shaped to egg-shaped, longer than wide and coarsely toothed with short, wide teeth, unlobed and with a U-shaped basal sinus and lower surfaces mostly hairless except on the leaf veins and in their angles. Leaves bright green above and pale green below. Flowers small, yellow-green, in compound branched clusters opposite the leaves. Fruits bluish, glaucous, sweetening and becoming palatable following frost in late fall.

Distribution: Of woods, wood edges, fields, roadsides, and trees along rivers and streams. Throughout the eastern United States through the eastern third of Texas.

Field recognition: Sweet or graybark grape: Leaves that are often wider than long and more distinctly lobed with a dull green upper surface and a pale green lower surface with cobwebby or woolly hairs. Fruit black or purplish with bloom, finally sweet. The small bronzy flowers hang in clusters that are eventually replaced by the more familiar purple fruits.

Similar species: Raccoon-grape (*Ampelopsis cordata* A. Michaux), a vegetatively similar member of the grape family, may be distinguished from true grapes (*Vitis* spp.) by its bark, which is tight rather than loose and shreddy as in grapes. The leaf blades are heart-shaped to triangular egg-shaped, sometimes shallowly lobed, with coarse, sharp teeth, and are always smooth and hairless (glabrous). In addition, the small, greenish flowers have free petals that fall separately, whereas those of grapes are united at their tips, falling as a cap from the flowers. Raccoon-grape grows in moist bottomland woods throughout the southeastern quarter of the United States, west through Central Texas, to the Texas Panhandle and Mexico.

Distribution: Sweet grape grows in wetlands along stream banks, river and creek banks, bottomlands, and pond margins throughout the southeastern United States west through East and north-central Texas to its southern margin in the Central Texas wetlands.

Field recognition: Mustang grape: Easily distinguished by the thick white wool that obscures the undersurface of its dark green, usually unlobed leaves (leaves on juvenile sucker shoots may be deeply lobed). It flowers and fruits earlier than the other two species, and by mid- to late summer its large, acidic fruits may be seen littering the trails far below the floodplain canopy in Palmetto State Park, where it grows as a high-climbing liana.

Distribution: Mustang grape ranges throughout the eastern half of Texas and adjacent regions of Louisiana, Arkansas, and Oklahoma.

PARASITES

Dodder
Unidentified *Cuscuta* sp.
CUSCUTACEAE (DODDER FAMILY)
Fig. 3-35
Flowering summer–fall

Field recognition: Yellowish or orange, threadlike, parasitic annual herbs lacking leaves and roots, and twining over foliage and stems of host plants, attached by adherent and penetrating bumplike haustoria. Hosts often members of the sunflower family (Asteraceae) in open, moist areas such as the Ottine Wetlands. Flowers small and numerous in small, branched clusters, pale whitish or yellowish. Fruit a capsule. Of open, moist or wet areas such as seepage slopes and pond margins where host plants occur, low open ground, and stream bottoms.

Similar species: Six species of dodder have been reported from the area (Turner et al. 2003a): *C. cuspidata* G. Engelmann, *C. glomerata* J. Choisy (Parks 1935a), *C. gronovii* C. von Willldenow *ex* J. A. Schultes, *C. indecora* J. Choisy, *C. obtusiflora* K. Kunth, and *C. pentagona* G. Engelmann. All are parasites of herbaceous plants of low open ground and stream bottoms. The last-mentioned species is a frequent parasite of cultivated legumes such as clovers and alfalfa (Diggs, Lipscomb, and O'Kennon 1999). Other hosts are members of the sunflower family, water-willow (*Justicia americana*), *Polygonum* spp., and *Euphorbia* spp., among many others.

Remarks: Dodder is a leafless, rootless holoparasite that lacks the green pigment chlorophyll and the consequent ability to manufacture its own food or to

Fig. 3-35
Dodder (unidentified
Cuscuta sp.) on
climbing boneset
(*Mikania scandens*).

take up water and inorganic nutrients from the soil, relying on its attachment
to other plants for both. Seedlings are reported to creep wormlike along the
soil surface after germination, growing at one end and shriveling at the other,
until they either reach a suitable host or exhaust their modest resources and
die trying. Under favorable conditions they may persist up to four weeks in
such a state (Holm et al. 1997). Undoubtedly the great majority perish in the
attempt. Success results in an eerie orange, spaghetti-like vine that is difficult to identify to the species level (Diggs, Lipscomb, and O'Kennon 1999).
The unidentified specimen in the photograph, shown parasitizing climbing
boneset (*Mikania scandens*) in the South Soefje Swamp, is offered as a representative of the genus. In the nearby Lost Pines forest we encountered and
identified *C. cuspidata*.

Distribution: These species tend to be widely distributed throughout Texas and
the central and eastern United States.

4
Grasses and Grasslike Plants

Grasses and their relatives, the sedges, rushes, and cattails, constitute a distinctive and easily recognized component of the wetland flora. They are characterized by green tubular or triangular stems and telescoping leaves that consist of a basal, tubular, sheathing portion and a terminal, free blade that, when present, is usually long and narrow with many close, parallel veins. This bladelike portion may be lacking in some sedges and rushes, however. These stems and leaves elongate and grow via basal and internodal meristems (regions of cell division), often from at or below ground level. Depending on their growth and branching behavior, they may form dense tussocks (as in "bunch grasses"), or spread widely, forming turfs. They are typically wind pollinated and consequently bear large numbers of reduced, nonshowy flowers in branched panicles, spikes, or dense, headlike clusters. **Grasses** (family Poaceae) have round or flattened stems (culms) with joints (nodes) and leaves in two alternating rows with the sheathing portion split lengthwise on the side opposite the blade. Each reduced flower (floret) is surrounded below by two scalelike, reduced leaves (bracts). These, in turn, are clustered into earlike units known as spikelets. **Sedges** (family Cyperaceae) have usually triangular, though sometimes round, stems without joints, and leaves in three rows with their tubular sheaths continuous around the stem, the bladelike portion sometimes lacking entirely. The reduced flowers are arranged in conelike spikes, each surrounded below by a single scale. **Rushes** (family Juncaceae) may be grasslike or sedgelike vegetatively, though never with triangular stems. However, their flowers and fruits are distinctive. The flowers, though reduced, are radially symmetrical and possess six dry, scaly, bractlike perianth parts (tepals, or undifferentiated petals/sepals) that surround the tiny, superior ovary like a tiny lily flower. Moreover, the fruits are dry capsules containing many tiny seeds rather than hard, dry, single-seeded achenes as in the grasses and sedges. **Cattails** (family Typhaceae) are unique and immediately recognizable by their marsh habitat; large, soft-spongy leaves in two rows; and dense, felty, cylindrical, 2-parted spikes.

Grasses may be subdivided into subxeric, drought-adapted species associated with uplands, and mesic or wetland (hydric) species associated with lowland seepage slopes, peaty wetlands, and floodplains. There is also an adventive, weedy component occurring primarily in disturbed areas, especially roadsides, agricultural land, and pastures. Upland grasses consist of species from the neighboring Blackland Prairie (tall to midgrass) such as silver bluestem (*Bothriochloa laguroides*) and purple threeawn (*Aristida purpurea*), as well as short-grass species from the arid Edward's Plateau and farther west, such as buffalo grass (*Buchloe dactyloides*). Upland grasses show a strong central and western prairie influence.

Lowland, moisture-loving grasses are of eastern U.S., southeastern Coastal Plain, or eastern Tallgrass Prairie affinity. Characteristic lowland grasses of eastern affinity are inland sea-oats (*Chasmanthium latifolium*), barnyard grass (*Echinochloa crus-galli*), Canada wild-rye (*Elymus canadensis*), Virginia wild-rye (*E. virginicus*), satin grass (*Muhlenbergia schreberi*), beaked panic grass (*Panicum anceps*), paired rosette grass (*P. (Dichanthelium) dichotomum*), variable rosette grass (*P. (D.) divergens*), velvet rosette grass (*P. (D.) scoparium*), and knot-root bristle grass (*Setaria parviflora*).

Lowland species of southeastern Coastal Plain affinity include giant reed (*Arundo donax*, adventive), carpet grass (*Axonopus fissifolius*), basketgrass (*Oplismenus hirtellus*), Timothy canary grass (*Phalaris angusta*), sugarcane plume grass (*Saccharum giganteum*), American cupscale (*Sacciolepis striata*), Gulf cordgrass (*Spartina spartinae*), and giant cutgrass (*Zizaniopsis miliacea*).

Lowland grasses with an eastern Tallgrass Prairie affinity include bushy bluestem (*Andropogon glomeratus*), broomsedge (*A. virginicus*), and switchgrass (*Panicum virgatum*). Again, these lowland species are of either eastern U.S., southeastern Coastal Plain, or eastern Tallgrass Prairie affinity and lend further support to the hypothesized relict nature of this ecosystem. By contrast, uplands show a strong central and western prairie component.

Common weedy, mostly adventives include rescue grass (*Bromus catharticus*), Japanese brome (*B. japonicus*), Bermuda grass (*Cynodon dactylon*), hairy crab grass (*Digitaria sanguinalis*), jungle-rice (*Echinochloa colona*), barnyard grass (*E. crus-galli*), little barley (*Hordeum pusillum*, native), rabbit's-foot grass (*Polypogon monspeliensis*), and Johnson grass (*Sorghum halapense*).

At the source of the San Marcos River in San Marcos, Texas, but found nowhere else in the world, is Texas wild rice (*Zizania texana*) (Mohlenbrock 2002). The species is restricted to clear, cool, fast-flowing spring waters located 50 km upstream from our study sites and may have been harvested for food by American Indians that once lived there. According to one source this endemic species is becoming rarer (Correll and Johnston 1970). We saw no Texas wild rice in the Ottine Wetlands.

Sedges of these ecosystems are nearly all eastern or southeastern wetland species at or near their western limits, with a few species of south-central or widespread southern distribution in North America. Distinctive wetland (hydric) species include crowfoot caric sedge (*Carex crus-corvi*), sallow caric sedge (*C. lurida*), sheathed flat sedge (*Cyperus haspan*), umbrella sedge (*C. involucratus*), false nut-grass (*C. strigosus*), tropical flat sedge (*C. surinamensis*), septate spike sedge (*Eleocharis interstincta*), large spike sedge (*E. palustris*), slender spike sedge

(*E. tenuis*), hairy fimbristylis (*Fimbristylis puberula*), western umbrella sedge (*Fuirena simplex*), white-top umbrella-grass (*Rhynchospora* (*Dichromena*) *colorata*), cluster beak-rush (*R. glomerata*), three-square bulrush (*Scirpus* (*Schoenoplectus*) *americanus*), and woolly-grass bulrush (*S. cyperinus*). Examples of subxeric to mesic upland species that infiltrate here are cedar sedge (*Carex planostachys*), reflexed sedge (*C. retroflexa*), Baldwin flat sedge (*Cyperus croceus*), one-flower flat sedge (*C. retroflexus*), and common or small-flowered hemicarpha (*Lipocarpha micrantha*).

Rushes show a similar eastern, southeastern, and central distribution pattern, with one widespread species (*Juncus effusus*). Slimpod rush (*J. diffusissimus*), soft rush (*J. effusus*), and roundhead rush (*J. validus*) are typical wetland or floodplain species, whereas inland rush (*J. interior*) and path rush (*J. tenuis*) are subxeric (seasonally moist) to mesic species.

Cattails offer an interesting case of the possible replacement or absorption (via hybridization) of one species (*Typha latifolia*) by another (*T. domingensis*). Early reports cite the former species (Bogusch 1928; Parks 1935a), whereas only the latter has been reported recently (Williams and Watson 1978). We also observed only the latter species. The common cattail (*T. latifolia*) is widespread throughout North America, whereas southern tule or giant cattail (*T. domingensis*) ranges across the southern United States to tropical America.

GRASSES

Wood-oats
Chasmanthium latifolium
(A. Michaux) H. Yates
POACEAE (GRASS FAMILY);
CENTOTHECEAE (CENTOTHECA TRIBE)
Fig. 4-1
Flowering June–September

Field recognition: Short-rhizomatous perennial grass of mesic woodlands and floodplains with broad leaves and conspicuous branched inflorescences of large, drooping, laterally flattened, wide "ear"-like spikelets of a dozen or so fertile florets and a sterile lower floret. An excellent field "primer" of grass spikelet morphology for the purposes of instruction or review.
Similar species: None.
Remarks: The flat, drooping, green spikelets of perennial wood-oats,

Fig. 4-1
Wood-oats (*Chasmanthium latifolium*) in flower.

also known as inland sea-oats, may be seen in summer along roadsides and trails. Despite its name, it is not cultivated for food. Preferred habitats include moist wooded areas and seasonally moist depressions where water collects in semi-arid forest.

Distribution: Throughout the southeastern United States south and west through the eastern half of Texas to south-central Texas.

Virginia wild-rye
Elymus virginicus C. Linnaeus
POACEAE (GRASS FAMILY); TRITICEAE (WHEAT TRIBE)
Fig. 4-2
Flowering (April–)May–August

Field recognition: Perennial bunch grass with erect culms terminating in dense, erect spikes of relatively short- and stiff-awned spikelets, suggesting an "ear" of wheat.

Similar species: Canada wild-rye (*E. canadensis* C. Linnaeus) is a close relative of similar habitats throughout temperate North America and Texas except the extreme southeast. However, it is easily distinguished by the thicker, inclined, or slightly nodding spikes composed of spikelets bearing long, thin, curved awns rather than short, stiff ones.

Remarks: The common name and appearance suggest that this perennial plant might be cultivated as a cereal grain but, like wood-oats, is not. The two are also similar enough in their habitat requirements to be found growing alongside one another in the public wetlands.

Distribution: Widespread in the United States except the southwestern and western half of Texas.

Fig. 4-2
Virginia wild-rye
(*Elymus virginicus*)
in flower.

Bristle basketgrass
Oplismenus hirtellus (C. Linnaeus)
A. Palisot de Beauvois
POACEAE (GRASS FAMILY);
PANICEAE (MILLET TRIBE)
Fig. 4-3
Flowering summer–October

Field recognition: Perennial grass with creeping culms rooting at the nodes. Leaf blades dark green, short and broad, often appressed against the substrate, suggesting the growth habit of wandering jew (*Zebrina* sp.; Commelinaceae). Erect stems bearing few, scattered spikelets near their tips. Reduced leaves (glumes) containing the spikelets conspicuously awned, with bristlelike tips, the first longer than the second. Exserted, plume-like stigmas (pollen-reception organs) of fertile upper floret dark reddish-violet. Exserted anthers (pollen-producing structures) yellow.

Fig. 4-3
Bristle basketgrass (*Oplismenus hirtellus*).

Similar species: None.

Remarks: Broad, tapering leaves suggest something other than a grass to the nonbotanist. In fact, the cultivar *O. h. variegatus* is cultivated as a hanging basket plant, as is the look-alike wandering jew. This perennial plant requires shade and prefers moist soils like those offered by wetlands and shady, mesic forests.

Distribution: From the southeastern U.S. Coastal Plain through East, Southeast, and South Texas; south to Argentina.

Variable rosette grass
Panicum (Dichanthelium F. Gould) *divergens* K. Kunth
POACEAE (GRASS FAMILY); PANICEAE (MILLET TRIBE)
Fig. 4-4
Flowering April–June and again late summer–fall

Field recognition: Perennial, clumped grass proliferating from an overwintering rosette of frost-hardy basal leaves by leaning to sprawling culms bearing short, broad, dark green leaf blades that are heart-shaped at their bases (juncture with sheath). Producing exserted, open-branched inflorescences of chasmogamous (open—releasing and receiving pollen), reverse egg-shaped, milletlike spikelets in spring and barely exserted, contracted inflorescences

Fig. 4-4
Variable rosette grass (*Dichanthelium* (*Panicum*) *divergens*).

of cleistogamous (closed—remaining in bud) spikelets during late summer and fall.

Similar species: A much taller rosette grass of wetter habitats is velvet rosette grass (*P.* (*D.*) *scoparium* J. de Lamarck) (Fig. 4-5). In addition, it has smaller, hairier leaves and erect, robust culms (to 1.5 m tall) with swollen nodes and leaf sheaths clothed with dense, velvety, white hairs. Though it is at the southwestern limit of its range in the Central Texas wetlands, we found it growing in open areas of seepy muckland. It produces chasmogamous spikelets from May to August and cleistogamous spikelets from July to October. Velvet rosette grass occurs in moist, sandy soils throughout the southeastern United States and West Indies to a southwestern limit here.

Remarks: Variable rosette grass, often known as *P. commutatum* J. A. Schultes, is a perennial denizen of sandy woods in regions associated with rivers and other bodies of water.

Distribution: Eastern United States to South America with its southwestern U.S. limit here. It occurs from the eastern United States to Texas.

Fig. 4-5
Velvet rosette grass (*Dichanthelium* (*Panicum*) *scoparium*).

Switchgrass
Panicum virgatum C. Linnaeus
POACEAE (GRASS FAMILY); PANICEAE (MILLET TRIBE)
Fig. 4-6
Following August–November

Field recognition: Large rhizomatous, perennial grass, yet often forming large clumps in low, moist areas. Tall culms (to 3 m), with narrow (3–15 mm wide), flat blades and topped with large, diffuse, open-branching inflorescences of nearly stalkless, gaping, lance-shaped, beaked spikelets bearing a lower staminate (pollen-bearing) floret and an upper fertile (seed-bearing) floret.

Similar species: None.

Remarks: This perennial may reach heights of nearly 3 m. It prefers moist soil and was seen in its characteristically large clumps in the cordgrass marsh of Palmetto State Park. Switchgrass is a remnant species of the original Tallgrass Prairie, along with big bluestem (*Andropogon gerardii*), little bluestem (*Schizachyrium scoparium*), and Indian grass (*Sorghastrum nutans*) (Diggs, Lipscomb, and O'Kennon 1999).

Distribution: Throughout the eastern United States and southern Canada east of the Rocky Mountains, and throughout Texas.

Fig. 4-6
Switchgrass (*Panicum virgatum*) flowers.

Common reed
Phragmites australis (A. Canavilles)
K. von Trinius *ex* E. von Steudel
POACEAE (GRASS FAMILY); ARUNDINEAE (REED TRIBE)
Fig. 4-7
Flowering July–November

Field recognition: Very tall (to over 4 m), rhizomatous perennial grass with culms topped by terminal, relatively loose, open-branched clusters of tawny (straw-colored, purplish when young) hairy spikelets. The hairs arise from the spikelet axis (rachilla) to which the individual florets are attached. Of wet areas, ponds, marshes, roadside ditches, usually in tight, clayey soils.

Similar species: Giant reed (*Arundo donax* C. Linnaeus; Arundineae), an extremely tall (to 5.5 m or more), rhizomatous, perennial grass native to the Mediterranean region and widely cultivated (and adventive) throughout the southern United States and Texas, is recently reported from these wetlands (Williams and Watson 1978; Fleenor and Taber, pers. obs.). It is much taller with wider (4–7 cm versus 1.5–6.0 cm), pale, glaucous-green leaves that alternate in two distinct rows on opposite sides of the culms (distichous) that are topped with erect, elongate, densely hairy, silvery-purplish to buffy, compact, spindle-shaped, branched inflorescences. Unlike those in common reed, these hairs arise not from the stalk (rachilla) to which the individual florets are attached but from the base and backs of the outer scales (lemmas) that surround the florets themselves. Giant reed is apparently sterile, spreading vegetatively and via cultivation. Of tight clay soils in wet areas and roadsides.

Remarks: Common reed is a tall, perennial, semi-aquatic grass, also known as "Danube grass." We saw it at the oxbow lake in Palmetto State Park and along the edge of the Gulf cordgrass marsh nearby, where it forms dense stands that are difficult to penetrate.

Distribution: Cosmopolitan in temperate and tropical regions throughout the world, it has one of the widest distributions of any plant or animal in the wetlands, occurring on all continents except Antarctica. Needless to say, it occurs throughout the state of Texas.

Fig. 4-7
A thick stand of common reed (*Phragmites australis*).

American cupscale grass
Sacciolepis striata (C. Linnaeus) G. Nash
POACEAE (GRASS FAMILY); PANICEAE (MILLET
TRIBE)
Fig. 4-8
Flowering August–November

Field recognition: A medium to
tall, rhizomatous perennial grass
of moist, sandy soils of streams,
marshes, and bogs, reaching its
southwestern limit here (Turner et
al. 2003b). We found it in floating mats of vegetation and muck
at the edge of the North Soefje
Swamp. Its leaves are conspicuously longitudinally parallel-veined
with culms terminating in spikelike
inflorescences of loosely appressed,
ascending branches bearing hairless, green spikelets on very short
stalks. The base of each spikelet
is surrounded by a conspicuously bulging, grooved or nerved,
reduced leaf (glume), hence the
common name "cupscale."

Fig. 4-8
Flowers of American cupscale grass
(*Sacciolepis striata*).

Similar species: None.

Remarks: Cupscale is a perennial grass of moist, sandy areas and may often be
found growing along the edge of a pond or in a marsh. It reaches a height of
nearly 1.5 m.

Distribution: From the Atlantic and Gulf Coastal Plain states to south-central
Texas, where it is rare (Correll and Johnston 1970). It reaches its southwestern
limit in the Ottine Wetlands.

Gulf cordgrass
Spartina spartinae (K. von Trinius) E. Merrill *ex* A. Hitchcock
POACEAE (GRASS FAMILY); CYNODONTEAE (BERMUDA GRASS TRIBE)
Fig. 4-9
Flowering spring–summer (fall)

Field recognition: Robust, tufted perennial grass in large bunches or tussocks.
Leaves dark green and tightly inrolled along their entire length, appearing
tubular or cylindrical with sharp, spinelike tips. Inflorescence a narrow, spikelike cluster of many shorter, appressed, overlapping, 1-sided spikes of individual single-flowered spikelets. Spikelets unawned.

Similar species: Prairie cordgrass
(*S. pectinata* J. Link), reported by
both Bogusch (1928) and Parks
(1935a), is a widespread species of
low, moist areas throughout the
northeastern and north-central
United States and southern Canada
and is at or near the southern limit
of its range here. It is a robust,
rhizomatous grass with flat leaves
and inflorescence spikes composed
of 1-sided spikes diverging slightly
from the axis, each with many conspicuously awned spikelets.

Remarks: This is a perennial, tough,
sharp-pointed grass that grows in
distinctive clumps and forms the
dominant vegetation in a single
marsh near the oxbow lake. We
saw only scattered individuals
elsewhere. The soil of the marsh is
slightly saline as might be expected
for a species with affinities more
coastal than inland. We discovered
that the water in which it grows is

Fig. 4-9
Gulf cordgrass (*Spartina spartinae*) flowers
(see also Fig. 1-24).

also more alkaline than any other water we tested (pH 8.0). Elsewhere these
wetlands tend to be acidic or neutral. An irrigation pipe leading from a water
source at the adjacent Warm Springs Foundation attests to the state's interest
in maintaining the cordgrass marsh whether or not natural conditions are
favorable for its survival.

Distribution: This is a Gulf and Caribbean coastal saltmarsh species of the
United States, Mexico, Paraguay, and Argentina. It occurs as a coastal disjunct in scattered inland localities, the Ottine Wetlands being one example.

Giant cutgrass
Zizaniopsis miliacea (A. Michaux) J. Doll & P. Ascherson
POACEAE (GRASS FAMILY); ORYZEAE (RICE TRIBE)
Fig. 4-10
Flowering May–September

Field recognition: Tall, coarse, rhizomatous perennial grass (to over 3 m) of
marshes, creek bottoms, and lakeshores, often forming extensive beds in shallow water or wet soil. Leaf blades wide (to 3.5 cm), pale green with light yellow
midveins and finely saw-toothed, cutting margins. Inflorescences large, loose,
open-branched clusters of narrow, single-flowered, unisexual spikelets.

Similar species: None.

Remarks: This aquatic grass is easily mistaken for cattail from a distance. We found it most abundant in North Soefje Marsh, where it suggests a slice of the Florida Everglades. Giant cutgrass is a sharp-edged perennial that may grow as tall as 3 m, and its alternative common name of "southern wild rice" indicates its close relationship to an edible species of the same botanical tribe.

Distribution: Throughout the southeastern United States south and west through East and Southeast Texas to the Edwards Plateau.

Fig. 4-10
Giant cutgrass (*Zizaniopsis miliacea*) in flower.

SEDGES

Crowfoot caric sedge
Carex crus-corvi R. Shuttleworth
ex G. Kunze
CYPERACEAE (SEDGE FAMILY)
Fig. 4-11
Fruiting late winter–early summer

Field recognition: Coarse, robust, densely bunched (short-rhizomatous) sedge of seasonally inundated floodplain bottomlands and openings with broad (to 12 mm wide), flat, glaucous-green leaf blades to nearly 1 m long. Culms terminating in contracted, plume-like inflorescences with numerous branches composed of short spikes with staminate (male) flowers at the top and pistillate (female) flowers at the base. Seed-sacs 6–8 mm long with a long beak and a broadly flattened triangular to heart-shaped body containing a flattened, lentil-shaped seed.

Fig. 4-11
Crowfoot caric sedge (*Carex crus-corvi*) in flower.

Similar species: Hyaline-scale caric sedge (*C. hyalinolepis* E. von Steudel) is a species of similar habitats, especially black clayey or mucky soils that, because of its coarse, robust habit and glaucous-green leaf blades with sharp, finely saw-toothed edges (4–13 mm wide), looks vegetatively quite similar to crowfoot caric sedge. However, it may be distinguished by its colonial, rhizomatous habit, spreading via long, creeping rhizomes as well as by its inflorescences, which are nonbranching and composed of separate staminate and pistillate spikes, the former terminal and erect, and the latter lower down, inclined outward. Hyaline-scale caric sedge is a species of the eastern half of the United States and is at the southwestern limit of its range here.

Remarks: This perennial sedge is common on muddy ground of seasonally inundated floodplain terraces in Palmetto State Park and is generally associated with marshes, swamps, and other seasonally saturated wetlands.

Distribution: Crowfoot caric sedge occurs in the midwestern Mississippi River drainage and southeastern Coastal Plain west to the southwestern margin of its range here.

Bayard Long caric sedge
Carex longii K. Mackenzie
CYPERACEAE (SEDGE FAMILY)
Fig. 4-12
Fruiting May–November

Field recognition: Densely clumped perennial sedge of wet sandy or peaty soils in open sites. Culms terminating in short clusters of few to several small, ovate spikes with staminate flowers around the base and pistillate flowers bearing seed-sacs (perigynia) distally, composing the bulk of the spike. Perigynia thin, flat, reverse egg-shaped with a short beak.

Similar species: Two closely related (also belonging to *Carex* section Ovales), very similar species are kidney-shaped caric sedge (*C. reniformis* (L. Bailey)

Fig. 4-12
Flowers of Bayard Long caric sedge (*Carex longii*).

J. K. Small) of shaded, wet woods and floodplains or bottomlands; and short caric sedge (*C. brevior* (C. Dewey) K. Mackenzie *ex* J. Lunell) of drier to submesic open, sunny habitats of calcareous, circumneutral meadows, prairies, and roadsides. Both may be distinguished from Bayard Long caric sedge by their broader, rounder (rounded versus reverse egg-shaped, about as wide as or wider than long versus longer than wide) seed-sacs. Apart from habitat, they may be distinguished by the wider than long, kidney-shaped seed-sacs of *C. reniformis* versus the about as wide as long, rounded seed-sacs of *C. brevior*. Kidney-shaped caric sedge is a species of the southeastern United States that reaches the southwestern margin of its range in or near these wetlands, whereas short caric sedge is widespread, often weedy and adventive throughout much of temperate North America and northern Mexico, but in Texas occurs mostly in the eastern quarter to third of the state.

Remarks: Bayard Long caric sedge is a perennial that prefers full sun rather than shade and inundated soils that may be peaty, sandy, or clayey.

Distribution: Bayard Long caric sedge is uncommon in Texas (Correll and Johnston 1970) but occurs through the eastern United States south and west to a western limit or margin here in the Ottine Wetlands and south into Mexico.

Hop caric sedge
Carex lupulina G. H. Muhlenberg
ex C. von Willdenow
CYPERACEAE (SEDGE FAMILY)
Fig. 4-13
Fruiting April–October

Field recognition: Coarse, robust, densely clumped to short, creeping-rhizomatous perennial sedge of open swamps and marshes in acidic to neutral or calcareous soils. Terminal spike thin, chaffy, and staminate. Lower spikes thicker, green with long-beaked seed-sacs 11–19 mm long.

Similar species: This species may hybridize with the similar sallow caric sedge (*C. lurida*) (Fig. 4-14), resulting in intermediate, sterile progeny (Reznicek 2002). That species may be distinguished by its thinner pistillate spikes with shorter, beaked seed-sacs (only 7.0–9.5 mm long).

Fig. 4-13
Hop caric sedge (*Carex lupulina*) in flower.

Remarks: Hop caric sedge is a perennial inhabitant of swamps (Diggs, Lipscomb, and O'Kennon 1999). The specimen shown here was photographed in Rutledge Swamp, where the plant thrives in the acidic waters.

Distribution: Through the eastern half of the United States to a southwestern limit with disjunction in the Ottine Wetlands of Gonzales County, Texas.

Sallow caric sedge
Carex lurida G. Wahlenberg
CYPERACEAE (SEDGE FAMILY)
Fig. 4-14
Fruiting April–August

Field recognition: Coarse, robust, short-rhizomatous, bunch-forming sedge of open marshes and swamps with thick triangular culms and broad (4–13 mm wide), flat leaf blades. Terminal spike erect, thin, narrow, staminate (pollen-producing) (3–6 cm long). Lower spikes thicker (2.5–4.0[–6.0] cm long by [12–]15–22 mm thick), green, spreading or nodding, pistillate and bearing long-beaked perigynia (sacs enclosing ovaries and subsequent seeds) 7.0–9.5 mm long. A weedy species of ditches and moist, disturbed areas in mostly sandy, acidic soils. Possibly recently introduced here from farther east, as our report is the first.

Fig. 4-14
Sallow caric sedge (*Carex lurida*) growing in a marsh among cattails.

Similar species: This species may hybridize with hop caric sedge (*C. lupulina*) (Reznicek and Ford 2002) (Fig. 4-13 [p. 108]), a species of similar range and habitat that is also quite similar in appearance but may be distinguished by its longer, beaked seed-sacs (11–19 mm long versus 7.0–9.5 mm long). The two species may produce sterile hybrid progeny where they co-occur (Reznicek 2002).

Remarks: Sallow caric sedge is a perennial relative of the grass family that grows abundantly on wet ground alongside giant cutgrass in North Soefje Marsh and in the cattail marsh of Palmetto State Park.

Distribution: Throughout the eastern United States and Canada south and west to a limit here in Gonzales County, Texas.

Small-tooth caric sedge
Carex microdonta J. Torrey & W. Hooker
CYPERACEAE (SEDGE FAMILY)
Fig. 4-15
Fruiting late April–June

Fig. 4-15
Flowers of small-tooth caric sedge
(*Carex microdonta*).

Field recognition: Small, creeping-rhizomatous perennial sedge of submesic, open areas on calcareous substrates such as prairies, meadows, and seeps. The ascending leafy culms terminate in a staminate spike with short, green pistillate spikes bearing seed-sacs in the angles of green scales on their lower portions.

Similar species: Cherokee caric sedge (*C. cherokeensis* L. Schweinitz) is taller, more robust, and densely rhizomatous from thick, dark chestnut-brown to black subterranean stems (rhizomes) with many roots. It also has drooping pistillate spikes on flexuous peduncles rather than erect to ascending ones.

The Ottine Wetlands are also the southwestern limits of five similar species of a group of about twenty clump-forming sedges native to floodplain forests of the eastern United States (*Carex* section Griseae). They appear to be very closely related, having descended from a common ancestor relatively recently (monophyletic); where they co-occur, they appear to co-exist by dividing up their alluvial habitat in very subtle ways according to soil texture, moisture availability, pH, and light levels (exposure). Hybridization may occasionally occur, but the offspring are usually sterile. They are a challenge to identify, as the ranges of measurements of diagnostic characters such as color and height of culm bases; seed-sac length and width; length, breadth, and shape of seeds; and stipe lengths (the tiny stalk supporting the seed) overlap considerably. These species are amphibious caric sedge (*C. amphibola* E. von Steudel), globose caric sedge (*C. bulbostylis* K. Mackenzie (*C. amphibola* var. *globosa*)), wrinkle-fruit caric sedge (*C. corrugata* M. Fernald), flaccid-fruit caric sedge (*C. flaccosperma* C. Dewey), and inflated caric sedge (*C. grisea* G. Wahlenberg (*C. amphibola* var. *turgida*)). An interesting discussion and key to this group can be found in Naczi and Bryson (2002).

Remarks: Small-tooth caric sedge is a perennial that grows in wet clearings in soil that is usually calcium-rich rather than sandy and acidic.

Distribution: From the Deep South of the Mississippi River valley and southern Coastal Plain region southwest and scattered widely throughout Texas.

Jointed flat sedge, chintule
Cyperus articulatus C. Linnaeus
CYPERACEAE (SEDGE FAMILY)
Fig. 4-16
Flowering May–October

Field recognition: Perennial reed or bulrushlike flat sedge from creeping, scaly rhizomes to over 1.5 m in height. Leaves reduced to basal, bladeless sheaths. Culms dark green, rounded to blunt-triangular in cross section, hollow with transverse septa at intervals, apparent upon drying or when squeezing and running thumb and forefinger along culms. Flowers borne in elongate spikes arranged in long-stalked nodding to nearly stalkless, spreading clusters at the ends of the culms. Leaves surrounding the inflorescence very short (3–11 mm long). Fruits tiny, 3-sided achenes, each containing a single seed. Of moist to wet clay grasslands and meadows.

Similar species: This is the only flat sedge in the United States with a reed or bulrushlike habitus. The thin, flattened, elongate, 2-rowed spikes arranged in spreading clusters will distinguish it from bulrushes (*Scirpus* and *Schoenoplectus* spp.), which have shorter, thicker, conelike spikes with scales and florets spirally arranged.

Remarks: This is a rhizomatous species that forms colonies in seasonally or permanently wet to moist clay soils of open floodplains and meadows.

Distribution: From northern South America (Colombia and Brazil) through South and Southeast Texas to the Gulf states (Florida). A colony has been reported from the Dallas area in north-central Texas (Diggs, Lipscomb, and O'Kennon 1999).

Fig. 4-16
Jointed flat sedge
(*Cyperus articulatus*).

Fig. 4-17
Umbrella flat sedge
(*Cyperus involucratus*)
in flower.

Umbrella flat sedge
Cyperus involucratus C. Rottboll
CYPERACEAE (SEDGE FAMILY)
Fig. 4-17
Flowering early summer–fall

Field recognition: Large clump-forming perennial flat sedge with umbrella-shaped inflorescences surrounded by 10–25 very long (15–40 cm), broad (1–15 mm), dark green, leafy bracts spreading like the ribs of an umbrella. Widely cultivated and sometimes escaping in the warmer regions of the world.

Similar species: None.

Remarks: Umbrella sedge may approach 1.5 m in height. It is a perennial Old World native that inhabits wet ground and has escaped from cultivation in the United States.

Distribution: Originally native to east Africa and Madagascar, it is adventive in Florida, Louisiana, Texas, and California.

False nut-grass
Cyperus strigosus C. Linnaeus
CYPERACEAE (SEDGE FAMILY)
Fig. 4-18
Fruiting June–October

Field recognition: Robust, non-rhizomatous, tufted perennial flat sedge of wet, sandy soils and marshy areas, sometimes weedy. Culms triangular, basally swollen, and rooting from a bulblike base. Leaf blades flat, narrow (1–4 [rarely 8] mm wide) and pale green. Inflorescence umbrella-like, surrounded by an involucre of a few long, leafy bracts, with spreading stalks bearing loose ovoid to elongate, open spikes of compressed, linear-lance-shaped, straw-colored to pale brown spikes.

Similar species: Fragrant or rusty flat sedge (*C. odoratus* C. Linnaeus) (Fig. 4-19) is a similar pantropical and warm-temperate species of moist to wet habitats throughout Texas and the United States. However, its spikes are nearly cylindrical and round in cross section or only slightly flattened, and have smaller scales (1.5–2.5[–3.2] mm versus 3.2–4.5[–6.0] mm long), whereas those of false nut-grass are distinctly flattened.

Remarks: This sedge is a close relative of the species used by ancient Egyptians to make papyrus for writing purposes. It may exceed 1 m in height and is a perennial denizen of bogs and marshes.

Distribution: False nut-grass is widespread in moist habitats across the United States and southern Canada and the eastern half of Texas (limited westward by aridity).

Fig. 4-18
False nut-grass (*Cyperus strigosus*) flowers.

Fig. 4-19
Fragrant flat sedge (*Cyperus odoratus*).

Cluster beak-rush
Rhynchospora glomerata
(C. Linnaeus) M. H. Vahl
CYPERACEAE (SEDGE FAMILY)
Fig. 4-20
Flowering summer–fall

Field recognition: Clump-forming, non-rhizomatous perennial sedge to over 1.5 m tall, with triangular culms and flat, narrow leaf-blades (to 5–6 mm wide). Inflorescences terminal and axillary heads of rich red-brown lanceolate to lance-ellipsoid spikes 4.5–6.5 mm long in the angles of leafy bracts, each containing two fruits. Dry, single-seeded fruits (achenes) dark chestnut-brown with long grayish beaks enclosed by bristles arising from the achene base. Of moist to wet, sandy acidic areas such as meadows, swales, fens, flatwoods, and bogs.

Fig. 4-20
Cluster beak-rush (*Rhynchospora glomerata*) flowers.

Similar species: A similar species with which cluster beak-rush is known to intergrade is *R. capitellata* (A. Michaux) M. H. Vahl of acid, boggy ground in East Texas, the eastern United States, southern Canada, and also California and Oregon. That species has narrower leaf blades (2–3 mm wide) and thinner culms (1–2 mm versus 2.0–3.5 mm thick at base) and smoothly convex achenes rather than with a conspicuous bump.

Distribution: Cluster beak-rush ranges through the southeastern United States south and west through East and Southeast Texas to a limit in or near the Ottine Wetlands.

Sword-grass
Scirpus americanus C. Persoon
CYPERACEAE (SEDGE FAMILY)
Fig. 4-21
Flowering April–July

Field recognition: Rhizomatous perennial sedge with creeping subterranean stems and dark green culms sharply triangular in cross section. Inflorescence a cluster of 1–4 small, sessile (stalkless), narrow, lance-ovoid brown spikes appearing lateral due to the vertical continuation of the single, surrounding bract leaf.

Fig. 4-21
Sword-grass (*Scirpus americanus*) flowers.

Similar species: None.

Remarks: This perennial sedge, also known as "three-square bulrush," grows in low, moist ground and in marshes throughout much of the world's temperate zone. It is a tall species capable of growing to heights of 1.5 m or more and is sometimes known as *Schoenoplectus pungens* (M. H. Vahl) E. Palla.

Distribution: Throughout the eastern half of the United States, southern Canada, and Texas; also the Pacific coast and at scattered localities in between.

RUSHES

Soft rush
Juncus effusus C. Linnaeus
JUNCACEAE (RUSH FAMILY)
Fig. 4-22
Flowering late April–June

Field recognition: Densely clumped perennial rush from short, thick rhizomes. Stems round, dark green, in dense bunches or sheaves to 1.25 m tall. Leaves short basal sheaths lacking blades, chestnut-brown. Inflorescence a branched cluster of pale brownish-tan flowers appearing lateral near the tip of the culm, overtopped by the sharp, round, awl-like surrounding bract leaf. In shallow water and moist, sandy soil.

Similar species: Woolly-grass bulrush (*Scirpus cyperinus* (C. Linnaeus) K. Kunth), a sedge reported by Bogusch (1928) and Parks (1935a), may be distinguished by its rounded-triangular stems, well-developed leaf blades, and terminal, drooping compound cymes of terminal spikelets. It ranges through the eastern United States and southern Canada to East Texas; also Washington and Oregon.

Remarks: Soft rush is neither sedge nor grass but a member of a smaller, closely related family of plants. In soft rush, the flowers appear to protrude from

Fig. 4-22
Soft rush (*Juncus effusus*).

one side of the stem, but technically they are terminal (at the tip). It is only a modified leaf associated with the flowers that rises above them. The fruits are dry capsules containing many tiny seeds, rather than the dry, single-seeded achenes of grasses and sedges.

Distribution: Widely distributed in the moist and flooded sandy soils of temperate North America. In Texas it ranges through the eastern and southeastern portions of the state west to the East Cross Timbers, with a southwestern limit here.

CATTAILS

Southern tule
Typha domingensis C. Persoon
TYPHACEAE (CATTAIL FAMILY)
Figs. 4-23, 4-24
Flowering April–July

Field recognition: Large (nearly 3 m tall) colonial, rhizomatous perennial herb with elongate, light yellow-green, alternate leaves in two rows, sheathing one within another at the stem base. Inflorescence unique and diagnostic; a dense, 2-parted, pale whitish-brown, cylindrical spike with pollen-producing staminate flowers above and pistillate flowers below, separated by a gap of 1–4 cm.

Similar species: Common or broad-leaf cattail (*T. latifolia* C. Linnaeus) is not quite so tall (only about 1.5 m) with the thicker, cylindrical, lower pistillate portion of the inflorescence dark brown and touching the pale, thin, upper staminate portion. It flowers during (March–)April–June and ranges throughout most of North America to Mexico.

Remarks: Cattails are not abundant in these wetlands, and just as remarkable is the replacement, entire or ongoing, of the only species mentioned in earlier surveys, the native broad-leaf cattail, by southern or giant cattail, a larger

Fig. 4-23
Southern tule (*Typha domingensis*) near oxbow lake.

Fig. 4-24
Flower spike detail of Southern tule (*Typha domingensis*).

competitor that is best viewed as an introduced plant. In fact, we did not see cattails at any site that would be described as natural. The giant species reaches nearly twice the height of its smaller congener and grows in small numbers at the edge of the Gulf cordgrass marsh, at the oxbow lake, and in larger numbers at a pond on private property nearby, and on a small neighboring marsh in Palmetto State Park watered by that same pond.

Distribution: Southern tule is distributed across the lower half of the United States from coast to coast and south to the tropical Americas and the West Indies.

5
Wildflowers

DICOTS

Our wildflower coverage is aimed at wetland species, but included here are those attractive and interesting upland plants observed while we moved to and from the swamps and marshes. Visitors to the wetlands will inevitably encounter them and will naturally want to know their names and something of their lives.

For the purposes of our treatment we consider "wildflowers" to be those terrestrial, herbaceous flowering plants that are neither grasslike, twining, trailing, nor of great stature. First we deal with the dicotyledons, or "dicots": those having broad leaves with a netlike pattern of veins and flower parts in multiples of two, four, or five. The next section treats the lilylike monocotyledons, or "monocots": those species having narrow, elongate foliage with parallel veins and flower parts in multiples of three or six. As a rather artificial default or "catch-all" category, wildflowers comprise a great diversity of plant species of varying degrees of relationship and apparency. They nevertheless display the familiar biogeographic pattern we identified for the other floral components examined thus far. This pattern is a biogeographic westward infiltration by eastern species near the western limits of their ranges in moist lowlands, and the eastward infiltration by western plants near the eastern limits of their ranges in dry, subxeric uplands. The history of this ongoing, dynamic interface is apparent from numerous examples of local species extirpations and invasions, from long ago to the present day. The often disjunct occurrence of many eastern species coupled with postglacial climatic change during the last eighteen thousand years underscores the relict nature of these wetlands. They survive because of the Carrizo Sands that hold water and allow species that are representative of eastern, more humid climates to cling to life. Notable eastern species of this description include purple gerardia (*Agalinus purpurea*), southern hog-peanut (*Amphicarpaea bracteata*), Texas blue-star (*Amsonia ciliata* var. *texana*), creeping slimpod (*A. tabernaemontana*), sicklepod (*Arabis canadensis*), azure aster (*Aster oolentangiensis*), swamp sunflower (*Helianthus angustifolius*), angle-stem water-primrose (*Ludwigia leptocarpa*), wild bergamot (*Monarda fistulosa*), yellow wood-sorrel (*Oxalis lyonii*), Florida pellitory (*Parietaria floridana*), hairy phacelia (*Phacelia hirsuta*), tearvine (*Polygonum sagittatum*), tufted buttercup (*Ranunculus fascicularis*), meadow beauty (*Rhexia mariana*), cluster sanicle (*Sanicula odorata*), lizard's tail (*Saururus cernuus*), butterweed (*Senecio glabellus*), seaside goldenrod (*Solidago sempervirens*), and primrose-leaf violet (*Viola primulifolia*).

Wildflowers marginal or disjunct at the southern limit of their ranges include prairie phacelia (*Phacelia strictiflora*), the Central Texas rough phlox (*Phlox pilosa* subsp. *latisepala* (*P. villosissima*)), and the temperate-boreal North American disjunct bog-parsnip (*Sium suave*).

South Texas and Mexican species disjunct or marginal at their northern limits include hoary blackfoot-daisy (*Melampodium cinereum*) and yellow sanvitalia (*Sanvitalia ocymoides*).

Western wildflowers marginal or disjunct to the east in these wetlands are yellow rock-nettle (*Eucnide bartonioides*), slender bladderpod (*Lesquerella recurvata*), and fan-leaf vervain (*Verbena plicata*).

Local endemics include Wright's false mallow (*Malvastrum aurantiacum*) and Texas umbrellawort (*Tauschia texana*).

False mint
Dicliptera brachiata (F. Pursh) K. Sprengel
ACANTHACEAE (WILD-PETUNIA FAMILY)
Fig. 5-1
Flowering June–November

Field recognition: Low-spreading, branching perennial herb of low, moist, shady areas and wooded stream bottoms with hexagonal stems and opposite, stalked, smooth-margined, elongate egg-shaped to narrowly egg-shaped leaves. Flowers borne in axillary clusters enclosed by 2–4 reverse egg-shaped, green, reduced leaves (bracts). Flowers tubular, 2-lipped, purple or pinkish with the pollen-bearing stamens enclosed in the upper lip. Fruit an ovoid capsule containing 2–4 seeds.

Fig. 5-1
False mint (*Dicliptera brachiata*).

Fig. 5-2
Henbit (*Lamium amplexicaule*).

Similar species: Henbit (*Lamium amplexicaule* C. Linnaeus) (Fig. 5-2) is a true mint (family Lamiaceae) native to Europe, the Mediterranean, and the Middle East that is now widely naturalized throughout North America and Texas. A low, weedy annual or biennial herb with purple or violet bilabiate, tubular flowers that appears similar to false mint from a height of nearly 2 m but upon closer examination proves to differ in nearly all respects. Its stems are square; its leaves broad, rounded, and coarsely scalloped; and the flowers have a slender, erect tube and a helmetlike or hooded upper lip and a lobed lower lip. During spring (February–May and later, depending on moisture) outcrossing, open (chasmogamous) flowers are produced, but during late fall and winter (November–February) it bears self-pollinating flowers that remain closed (cleistogamous). It is not strictly a wetland plant, but we encountered it, as we did many other wildflowers shown here, as we moved down from the uplands into the swamps and marshes.

Remarks: The purple flowers of false mint appear in a variety of wet habitats, including land adjacent to the San Marcos River. Its stems are remarkable for their hexagonal shape.

Distribution: Through the southeastern United States west through East and Southeast Texas to the East Cross Timbers and Edwards Plateau.

Drummond's wild-petunia
Ruellia drummondiana (C. Nees von Esenbeck) A. Gray
ACANTHACEAE (ACANTHUS FAMILY)
Fig. 5-3
Flowering (June–)August–September(–October)

Field recognition: Rhizomatous perennial, finely hairy herb to 1 m in height, with angled stems and opposite egg-shaped to broadly elliptic, dark green leaves distinctly feather-veined. Flowers showy, funnel-shaped to trumpet-shaped, lavender with darker blue nectar "guides" in throat and borne in clusters in the angles of the upper leaves. Some flowers much shorter than others,

Fig. 5-3
Drummond's wild-petunia (*Ruellia drummondiana*).

Fig. 5-4
Violet wild-petunia (*Ruellia nudiflora*).

suggesting cleistogamy (self-pollination) (Correll and Johnston 1970). Fruit a dry, explosively splitting capsule about 1 cm long. Of shady, wooded, riparian areas and often rocky slopes along streams, in canyons, and on floodplains.

Similar species: Violet wild-petunia (*R. nudiflora* (G. Engelmann *ex* A. Gray) I. Urban) (Fig. 5-4) may be distinguished by flowers (April–May to throughout year) that are borne in open-branching, terminal, rather than axillary clusters (in the leaf angles) as in Drummond's wild-petunia. Violet wild-petunia also bears more oblong-elliptic and less egg-shaped leaves. It occurs in sandy, open woods from north-central Texas through the southern half of the state to northeastern Mexico. Low wild-petunia (*R. humilis* T. Nuttall) is a more upland species of the eastern half of Texas and the United States reported for the Ottine Wetlands by Williams and Watson (1978). It may easily be distinguished by its smaller, nearly stalkless, hairy, egg-shaped to broadly elliptic or lance-shaped leaves. Flowering April (May)–September(–October).

Remarks: Wild-petunias often produce smaller, cleistogamous (closed) flowers that are self-pollinated early in the season followed by larger, showy chasmogamous (open) flowers that are cross-pollinated by butterflies and bees later in the season. Cleistogamy may be a strategy for ensuring the production of at least some seed during years when pollinators are scarce. The chasmogamous flowers are open in the morning with the showy, funnel-shaped petal tube falling in the afternoon, thus lasting a single day. Seeds are often sticky when wet (mucilaginous) and are flung from explosively splitting capsules by small arms (jaculators) on which they develop.

Distribution: Drummond's wild-petunia is endemic to the eastern and southeastern Edward's Plateau from north-central to south-central Texas. It is at the southeastern limit of its range in the Ottine Wetlands.

Wild-chervil
Chaerophyllum tainturieri W. Hooker
APIACEAE (PARSLEY FAMILY)
Fig. 5-5
Flowering mid-March–April(–May)

Field recognition: Erect aromatic annual herb with finely divided, parselylike leaves. Flowers tiny, white, and borne in compound flat-topped clusters (umbels) of terminal, few-flowered umbellets (little umbels). Fruits distinctive, narrowly oblong, beaked, ribbed mericarps on pedicels to 1 cm long. Coriander-like, aromatic odor. Of stream bottoms, lowland alluvial thickets, moist prairies, roadsides, and weedy in disturbed areas.

Fig. 5-5
Wild-chervil (*Chaerophyllum tainturieri*).

Similar species: None.

Remarks: Wild-chervil's white flowers bloom in riparian thickets as well as in disturbed uplands. Unlike the ill-smelling leaves of beautyberry, these have a pleasing aroma when crushed or rubbed.

Distribution: Throughout the southeastern United States west to Arizona and throughout Texas, but most common in the eastern half.

Water-hemlock
Cicuta maculata C. Linnaeus
APIACEAE (PARSLEY FAMILY)
Fig. 5-6
Flowering late May–July(–September)

Field recognition: Tall, hairless, gray-green perennial herb from a tuberous base and clustered, fleshy-tuberous roots to over 2 m high. Leaves twice or thrice feather-compound, of lance-shaped, short- to long-pointed, coarsely saw-

Fig. 5-6
The flowers of poisonous water-hemlock (*Cicuta maculata*).

toothed or divided by sharp teeth. Flowers small, white to cream, and borne in large, showy, compound, flat-topped, umbrella-shaped clusters. Fruits oval to rounded, pairs of seeds (mericarps) with low, corky ribs. Of marshes and swamps, along streams, and in other wet places.

Similar species: Poison-hemlock (*Conium maculatum* C. Linnaeus), a native of Europe and Asia, is adventive in stream bottoms and other wet places in the southern half of Texas. Williams and Watson (1978) reported it for the wetlands. It is a biennial herb to over 3.5 m tall with a strong, unpleasant petroleum/rubber/cumin odor, purple-spotted stems, and finely feather-divided, fernlike foliage. Flowering May–June(–August). Water or bog-parsnip (*Sium suave* T. Walter), a circumboreal and north-temperate species of swamps, marshes, and bogs, found from Newfoundland and South Carolina west to British Columbia and California, occurs in Texas only in the Ottine Wetlands and a few neighboring wetlands in the south-central portion of the state. It is possibly a glacial relict here, a remnant of a north-temperate or boreal flora dating from the most recent glacial maximum about eighteen thousand years ago. It may be recognized by its longitudinally ridged (corrugated) stems and once feather-compound leaves of lance-shaped to linear, saw-toothed to cut, widely spaced leaflets. A perennial herb to 1.2 m from spindle-shaped, clustered roots. Flowers white, in compact, compound, flat-topped clusters (May–September).

Remarks: Water-hemlock is a deadly, white-flowered plant that grows in the muck of the deep swamp. An herb reaching heights of over 2 m, it is also known as "beaver-poison" and "spotted cowbane." These names underscore its reputation as the most poisonous herb in the United States. One bite of its tissues may kill a person, and children have been poisoned merely by blowing whistles carved from its stem (Diggs, Lipscomb, and O'Kennon 1999).

Distribution: Through the eastern half of the United States and Canada west to the Dakotas, and through East and Southeast Texas to the Rolling Plains and Edwards Plateau.

Umbrella water-pennywort
Hydrocotyle umbellata C. Linnaeus
APIACEAE (PARSLEY FAMILY)
Fig. 5-7
Flowering (April–)May–October

Field recognition: Smooth, hairless perennial wetland or aquatic plant with creeping stems or rooting stolons and round, scallop-margined leaves, with the leafstalks attached near their middle. Flowers tiny, yellow-green to white in umbrella-shaped clusters exceeding the leaves.

Similar species: Whorled water-pennywort (*H. verticillata* C. Thunberg) (Fig. 5-8) has flowers in whorls, forming an interrupted spike. It flowers from May to October and ranges from coast to coast throughout the southern United States, south through the West Indies, Mexico, and Central and South America. Bogusch (1928) and Parks (1935a) reported floating water-pennywort (*H. ranunculoides* C. von Linné). It differs from the previous two species in having deeply notched, 5- to 6-lobed leaves with a basal sinus at the point of attachment of the leafstalk. It flowers from April to October and ranges through the eastern United States and eastern third of Texas to a southwestern limit here in the central part of the state; also in Arizona and the Pacific coast south to Panama, Cuba, and South America.

Remarks: Water-pennyworts are the epitome of wetland plants. They grow in swamps and marshes in muck thick enough to swallow knee-high boots, and in fact, we used them as early warnings of such terrain. Their appearance suggests a forest of small water-lilies perched with their pads in the air rather than on the surface of a pond. The umbrella water-pennywort stand shown in the photograph was growing in Soefje Swamp, as was the specimen of its whorled relative, shown in the accompanying photograph, which bears tiny yellow-green or white flowers.

Fig. 5-7
A stand of umbrella water-pennywort (*Hydrocotyle umbellata*).

Fig. 5-8
Whorled water-
pennywort
(*Hydrocotyle
verticillata*).

Distribution: Umbrella water-pennywort ranges from the eastern United States and southern Canada through the eastern half of Texas south into the West Indies and Mexico as far south as South America; also in Oregon and California.

Thread-leaf mock bishop's weed
Ptilimnium capillaceum (A. Michaux) C. Rafinesque-Schmaltz
APIACEAE (PARSLEY FAMILY)
Fig. 5-9
Flowering late May–August

Field recognition: Slender, erect, cryptic annual herb with leaves finely feather-divided into threadlike divisions. Flowers tiny, white, and borne in compound, flat-topped, umbrella-like clusters (umbels) of few-flowered umbellets (small umbels). Fruits small, ovoid. Of wet areas.

Fig. 5-9
Thread-leaf mock
bishop's weed
(*Ptilimnium
capillaceum*).

Similar species: Nuttall's mock bishop's weed (*P. nuttallii*) (flowering April–July), a midwestern and east-central U.S. species ranging through East and Southeast Texas to the East Cross Timbers and Edwards Plateau and at its southern limit here, is similar but has less finely divided leaves, reduced leaves (bracts) surrounding the umbels undivided, and longer, recurved styles (0.5–1.5 mm versus 0.2–0.5 mm long). It grows in moist prairies, sandy or silty open ground, and along roadsides and in other moist places.

Remarks: This white-flowered bishop's weed species is a wetland plant of both brackish and fresh waters in marshes, swamps, and woodlands.

Distribution: Throughout the southeastern United States through East and Southeast Texas to north-central Texas and the Edwards Plateau to its southwestern limit in the Ottine Wetlands.

Canada sanicle
Sanicula canadensis C. Linnaeus
APIACEAE (PARSLEY FAMILY)
Fig. 5-10
Flowering April–June

Field recognition: Biennial herb, hairless with dark green, fan-divided (palmate), 3–5-parted leaves, the divisions reverse lance-shaped and coarsely toothed. Flowers tiny, white, sessile (stalkless), and in clusters of a few at the ends of elongate stalks that compose the primary rays of the umbrella-shaped inflorescence (an umbel). Ovaries and fruits burlike with many curved or hooked bristles. Common in shaded, moist woods and floodplains.

Similar species: Cluster sanicle (*S. odorata* (C. Rafinesque-Schmaltz) K. Pryer & L. Phillippe) (Fig. 5-11) is similar but differs in having flowers (April–May) with tiny yellow-green petals (versus white) that are longer (rather than shorter) than the lance-egg-shaped sepal-lobes (versus lance-shaped and spine-tipped) and a long-exserted style (longer versus shorter than the bristles) on the fruit. It is a rarer species of moist woods in the eastern half of Texas and the United States.

Fig. 5-10
Canada sanicle
(*Sanicula canadensis*).

Fig. 5-11
Cluster sanicle
(*Sanicula odorata*).

Remarks: Canada sanicle bears small white flowers and grows in moist woods. It is one of the species featured here that are not found in field guides to Texas wildflowers, presumably because its blooms are not showy enough to attract attention.

Distribution: Throughout the eastern half of Texas and the United States.

Texas umbrellawort
Tauschia texana D. von Schlectendahl
APIACEAE (PARSLEY FAMILY)
Fig. 5-12
Flowering February–June

Field recognition: Stemless, taprooted, hairless perennial herb with finely divided, fernlike, twice feather-compound, basal leaves. Inflorescences are compound, flat- or round-topped, umbrella-like clusters of tiny yellow flowers borne aloft on long stalks. Fruits slightly compressed, oval mericarps (paired achenes) separated by a deep groove and with fine ribs containing oil tubes on the surface. A denizen of wet woods and floodplain thickets.

Similar species: None.

Fig. 5-12
Texas umbrellawort
(*Tauschia texana*).

Remarks: Texas umbrellawort is a yellow-flowered endemic that occurs in only about one dozen counties in the southern Blackland Prairie and Gulf Coastal regions of Texas.

Distribution: If the atlas of Turner et al. (2003a) is any indication, this species of very limited distribution is more abundant here in Gonzales County than anywhere else in the world. We found it in its preferred habitat of wet woods inside Palmetto State Park and in the South Soefje Swamp along the north branch of Rutledge Creek.

Willow slimpod
Amsonia tabernaemontana T. Walter
APOCYNACEAE (OLEANDER FAMILY)
Fig. 5-13
Flowering (March–)April(–May)

Field recognition: Erect perennial herb from a woody rootstock or rhizome with hairless, smooth-margined, lance-shaped to oblong-elliptic leaves and milky sap. Flowers azure-blue, tubular with a star-shaped mouth in compact terminal clusters. Fruits distinctive pairs of long, slender, spreading pods joined at base and containing numerous naked seeds (lacking a tuft of hairs). Of moist, sandy soils along lakes, streams, and stream bottom thickets.

Fig. 5-13
Willow slimpod (*Amsonia tabernaemontana*).

Similar species: Texas slimpod (*A. ciliata* T. Walter var. *texana* (A. Gray) J. Coulter) was reported for these wetlands by Parks (1935a). It is a species of rocky, gravelly, limestone-derived soils and less commonly in sandy loams of cedar brakes and grasslands on the Blackland Prairie, Rolling Plains, and Edwards Plateau. It ranges throughout the southeastern United States west to Central Texas. In addition to habitat, it may be distinguished by much shorter and narrower linear or threadlike leaves only 0.5–17.0 mm wide versus 8–30 mm wide, and less than 5 (rarely to 8) cm long versus greater than 5 cm long.

Remarks: Willow slimpod bears light blue petals, the free portions of which are arranged radially, in a star-shaped pattern.

Distribution: This plant has a preference for sandy terrain near bodies of water. It occurs from Pennsylvania south to Georgia, south and west through East and Southeast Texas to the East Cross Timbers and its southwestern limit in the Ottine Wetlands.

Indian-hemp
Apocynum cannabinum C. Linnaeus
APOCYNACEAE (OLEANDER FAMILY)
Fig. 5-14
Flowering April–July(–August)

Field recognition: Perennial herb from horizontal roots to over 1 m tall with opposite, short-stalked to sessile, egg-shaped, oblong-elliptic or lance-shaped, smooth-margined leaves and with milky sap. Flowers small, 5-parted, white to yellow or greenish, petals bell- or urn-shaped. Fruits long (4–22 cm), slender, paired pods containing numerous seeds with tufts of long, silky hairs at one end. In moist or wet, usually sandy, sometimes gravelly or clayey soils in open lowland and mesic woodlands, in fields, or along streams and riverbanks.

Fig. 5-14
Indian-hemp
(*Apocynum
cannabinum*).

Similar species: None.

Remarks: A relative of the slimpods (*Amsonia* spp.) as evidenced by its milky sap and slender, podlike fruits.

Distribution: Throughout the United States and southern Canada and across Texas to a southern limit in the Ottine Wetlands and Big Bend region.

Heath aster

Aster (Symphyotrichum G. Nesom) *ericoides* C. Linnaeus

ASTERACEAE (DAISY FAMILY); ASTEREAE (ASTER TRIBE)

Fig. 5-15

Flowering September–October(–November)

Field recognition: Rhizomatous, many-branched perennial herb with main branches erect to ascending or reclining and bearing many short, ascending, flower-head-bearing, 1-sided branchlets. Leaves of branchlets tiny (2–3 mm long by 1 mm wide), those of main stems longer, linear to oblong (1–2 cm long), but usually withered by flowering time. Flower heads small, crowded, with white rays and short (3–4 mm long) recurved scales, each with a fringe of short, straight hairs. Of open, disturbed, usually upland areas.

Fig. 5-15
Heath aster (*Aster ericoides*).

Similar species: Side-flower aster (*A.* (*S.*) *lateriflorus* (C. Linnaeus) N. Britton) is another rhizomatous perennial aster with a similar growth habit (ascending to long-arching main branches bearing short, 1-sided, head-bearing branchlets) and small flower heads with white rays (September–October[–November]). It may be distinguished by its habitat, which includes moist, sandy or boggy areas (Correll and Johnston 1970) as well as well-drained uplands (Diggs, Lipscomb, and O'Kennon 1999), and much larger elliptic-reverse lance-shaped to linear stem leaves ([3–]5–10[–15] cm long by [0.2–]1.0–2.0 [–3.5] cm wide) that are persistent at flowering time. It occurs throughout the interior of the eastern United States and southeastern Canada to East Texas and is disjunct to its southwestern limit in Gonzales County (Turner et al. 2003a).

Remarks: The white flowers of heath aster may be seen in a wide variety of unshaded, usually upland habitats. Leaves usually disappear before the flowers make their display. It is also known as "prairie aster."

Distribution: This plant ranges through the interior eastern United States and southern Canada south and west throughout Texas to northern Mexico.

Texas broomweed
Gutierrezia texana (A. P. de Candolle) J. Torrey & A. Gray
ASTERACEAE (DAISY FAMILY); ASTEREAE (ASTER TRIBE)
Fig. 5-16
Flowering July–November

Field recognition: Shrubby annual herb to three-quarters of a meter tall with small, sticky, linear leaves sparsely distributed along spreading, branching, green stems. Flower heads small (3–4 mm high), terminating the short branchlets, with short, curling, yellow rays and greenish-yellow disk florets. A weed of disturbed areas throughout the state.

Similar species: None.

Fig. 5-16
Texas broomweed (*Gutierrezia texana*).

Remarks: Texas broomweed is a yellow-flowered plant of disturbed habitats that also goes by the scientific name *Xanthocephalum texanum* (A. P. de Candolle) L. Shinners. Unlike some of its shrubby, perennial relatives, this is an annual species and is known in common parlance as a "broomweed" rather than "snakeweed."

Distribution: Restricted to Oklahoma, Texas, and Mexico.

Seaside goldenrod
Solidago sempervirens C. Linnaeus
ASTERACEAE (DAISY FAMILY); ASTEREAE (ASTER TRIBE)
Fig. 5-17
Flowering late summer to fall

Field recognition: Rhizomatous, perennial herb of marshy, often brackish habitats such as pond margins, ditches, and swales, with smooth, hairless stems to over 2 m in height, and semi-succulent (somewhat fleshy), entire, narrow, linear-lance-shaped leaves; grasslike at base; stem leaves shorter to narrowly elliptic. Flowers small, yellow, and aggregated into heads in narrow, elongate 1-sided (secund), condensed, spikelike clusters.

Similar species: Rough-leaf goldenrod (*S. rugosa* P. Miller) (Fig. 5-18), an eastern species of moist, sandy soils as well as sandy or rocky uplands bordering the

Ottine Wetlands, is quite different in being roughly hairy and bearing toothed leaves that are slightly sandpapery above and hairy beneath. It is much smaller in stature (often dwarfed by drought), with dark green, minutely net-veined, engraved leaves, and short, 1-sided spikelike or branched inflorescences (flowering September–November). Unlike seaside goldenrod it is not a wetland species but one likely to be encountered in moist, sandy soils nearby. It ranges through the eastern United States and southern Canada west to the Midwest, south and west through Missouri, Arkansas, East and Southeast Texas to a southwestern limit with a disjunction here.

Fig. 5-17
Seaside goldenrod (*Solidago sempervirens*).

Remarks: Goldenrods are not generally confined to wetlands, but seaside goldenrod is often found in marshy, even saline habitats.

Distribution: Seaside goldenrod occurs from the Atlantic and Gulf Coastal Plain from Massachusetts to Veracruz, Mexico, and is disjunct inland to its limit in the Ottine Wetlands (Turner et al. 2003a).

Fig. 5-18
Rough-leaf goldenrod (*Solidago rugosa*).

Bull thistle
Cirsium horridulum A. Michaux
ASTERACEAE (DAISY FAMILY); CARDUEAE
(THISTLE TRIBE)
Fig. 5-19
Flowering March–May

Field recognition: Coarse, prickly biennial or winter annual herb with long lance-shaped to reverse lance-shaped, feather-lobed, spiny leaves to over 2 m tall but usually much shorter. Flower head large, terminal, usually solitary with long spiny leaves forming a conspicuous, basketlike "false" involucre (calyculum) about the true involucral scales (phyllaries) of the head. Florets all of the disk type, with purplish, rosy-lavender, yellow, or whitish radially symmetrical petal tubes.

Fig. 5-19
Bull thistle (*Cirsium horridulum*).

Similar species: Milk-thistle (*Silybum marianum* (C. Linnaeus) J. Gaertner) (Fig. 5-20) is a coarse biennial or annual (perennial?) herb with prickly, thistlelike foliage to nearly 2 m high. Leaves long, broad, prickly-lobed or toothed with stem leaves with earlike, clasping lobes at the base, green mottled with white, particularly near the veins. Flowers (May–June) borne in thistlelike heads of spiny involucral scales and rosy-purple florets. A native of the Mediterranean region widely adventive in pastures, along roadsides,

Fig. 5-20
Milk thistle (*Silybum marianum*).

and in low disturbed areas on the Edwards Plateau, north-central Texas, and elsewhere. Not a wetland species per se, and originally reported as "rare and local" in Texas (Correll and Johnston 1970), it is widely adventive on the Edwards Plateau, north-central Texas, and elsewhere today (Diggs, Lipscomb, and O'Kennon 1999). The milky sap provides an antidote to certain types of mushroom poisoning.

Remarks: The alternative common name "yellow thistle" is a reminder that, in Southeast Texas, the bull thistle's flower is not always lavender or purple. We found the formidably spined plant growing in areas that are only periodically flooded.

Distribution: Through the Atlantic and Gulf Coastal Plain from Maine to East and Southeast Texas.

Mist flower
Eupatorium (Conoclinum A. P. de Candolle)
coelestinum C. Linnaeus
ASTERACEAE (DAISY FAMILY); EUPATORIEAE (BONESET TRIBE)
Fig. 5-21
Flowering August–November

Field recognition: Rhizomatous, perennial herb with opposite, stalked, hairy, triangular, toothed leaves. Of moist wooded areas in sandy or calcareous soils. Flowers borne in umbrella-like, flat-topped clusters of heads terminating ascending stems from subterranean rhizomes. Flower heads rayless, composed entirely of blue or purplish-blue disk florets. Fruits blackish, 5-ribbed, single-seeded achenes crowned by a ring of bristles.

Similar species: Pink boneset (*E. incarnatum* T. Walter), reported by Parks (1935a), is of similar habitat and growth habit but is perennial from a fibrous-rooted crown and has florets with pink or lilac-tipped white petals rather

Fig. 5-21
Mist flower
(*Eupatorium
coelestinum*).

than blue. This is a species of Mexico north to Arizona and through South, Southeast, and East Texas to the southeastern United States. It occurs inland to north-central Texas and the Balcones Escarpment not far west of the Ottine Wetlands.

Remarks: Purple-petaled mist flower grows in moist wooded areas and is tolerant of calcareous soils in addition to the sandy soils more typical of the Ottine Wetlands.

Distribution: Throughout the southeastern United States west through East and Southeast Texas to north-central Texas and the Edwards Plateau.

Prostrate-sunflower, horse herb

Calyptocarpus vialis C. Lessing
ASTERACEAE (DAISY FAMILY); HELIANTHEAE (SUNFLOWER TRIBE)
Fig. 5-22
Flowering April–July (sporadically throughout the year)

Field recognition: Short trailing, creeping, or sprawling perennial herb with opposite egg-shaped to lance-egg-shaped or triangular, toothed, hairy leaves. Flower heads tiny, sunflower-like with a few to several short yellow, marginal rays and small yellow disks. Fruits tiny flattened achenes crowned by two pointed awns. A weed of moist, partly open to shady areas such as roadsides, trailsides, and lawns.

Similar species: None.

Remarks: Horse herb's yellow flowers grow in shaded areas and often appear as "noxious" weeds in lawns.

Distribution: Occurring widely from Central America north through Mexico to the Gulf Coastal Plain of the southeastern United States, including the southeastern two-thirds of Texas.

Fig. 5-22
Prostrate-sunflower
(*Calyptocarpus vialis*).

Lindheimer-daisy, yellow star-daisy
Lindheimera texana A. Gray & G. Engelmann
ASTERACEAE (DAISY FAMILY); HELIANTHEAE (SUNFLOWER TRIBE)
Fig. 5-23
Flowering March–May

Field recognition: Taprooted annual herb with a usually simple stem to 30 cm tall crowded with bright green, reverse egg-shaped to reverse lance-shaped, stalkless leaves that are coarsely toothed in their apical half. Flower heads solitary on short stalks. Marginal ray flowers pistillate, fertile (seed-bearing), 4–5 in number with yellow, 1 cm long, straplike petals having two apical teeth. Central disk flowers perfect but infertile, producing pollen only (functionally staminate or male) with short, 5-toothed yellow or brownish-yellow petal tubes. A prairie wildflower.

Similar species: Engelmann-daisy (*Engelmannia pinnatifida* A. Gray *ex* T. Nuttall), another monotypic, upland species, differs in being a taller, perennial, densely hairy herb with deeply feather-lobed leaves. Flower heads with eight pistillate ray florets ("little flowers") that are 3-toothed at their tips. An inhabitant of calcareous open uplands rather than sandylands throughout Texas, ranging from the Great Plains of the south-central United States to north-central Mexico.

Remarks: These wildflowers are not species that require wetland soils. They grow in upland pastures and may be seen at the interface between pastures and swamps. Both Lindheimer- and Engelmann-daisies are monotypic—the sole members of unique genera comprised of only a single species apiece.

Distribution: Nearly endemic in prairies of north-central and Central Texas, yellow star-daisy also occurs southward to Coahuila, northern Mexico.

Fig. 5-23
Lindheimer-daisy
(*Lindheimera texana*).

Mexican hat
Ratibida columnifera (T. Nuttall)
E. Wooton & P. Standley
ASTERACEAE (DAISY FAMILY);
HELIANTHEAE (SUNFLOWER TRIBE)
Fig. 5-24
Flowering May–October

Field recognition: Biennial or
perennial herb with mature leaves
feather-cleft into linear to lanceolate divisions. Flower heads
cylindrical, with 3–7 marginal rays
ranging in color from yellow with
a red-brown spot at the base to
entirely yellow or reddish-brown.
Of open, calcareous soils and disturbed areas.

Similar species: None.

Remarks: This yellow and red flower
was growing on the floodplain of
the San Marcos River in an area
that is dry much of the year. The
black beetles feeding on the flower heads are unidentified weevils.

Fig. 5-24
Mexican hat (*Ratibida columnifera*).

Distribution: Mexican hat occurs from the central U.S. plains, through the
western two-thirds of Texas, to north-central Mexico and is perhaps adventive eastward to the Mississippi River region.

Rough-stem or swamp rosinweed
Silphium radula T. Nuttall
ASTERACEAE (DAISY FAMILY);
HELIANTHEAE (SUNFLOWER TRIBE)
Fig. 5-25
Flowering June–July

Field recognition: Coarse perennial herb to over 2 m from a woody taproot
with very hairy, sandpapery, sessile, entire lance-egg-shaped leaves. Flower
heads large, sunflower-like with yellow marginal rays. Flattened, winged,
single-seeded achenes produced by the ray flowers; disk florets staminate
(shedding pollen only). Of seasonally wet, open areas in alluvial, floodplain
soils, calcareous or sandy, and along roadsides.

Similar species: Sunflowers (*Helianthus* spp.) are similar in appearance but
have fertile, seed-bearing disk flowers and infertile ray florets; in rosinweeds
(*Silphium* spp.) the reverse is the case.

Fig. 5-25
Rough-stem rosinweed
(*Silphium radula*).

Swamp sunflower (*H. angustifolius* C. Linnaeus) is a perennial herb of moist areas from a short, fibrous root crown to 1.5 m tall. It is mostly unbranched with linear to narrowly lance-shaped, stalkless, very rough-sandpapery leaves 1–20 cm long by 3–15 (rarely 20) mm wide. Inflorescences open-branched above. Flower heads small with yellow marginal rays and purple-brown discs about 1 cm in diameter (flowering late summer–fall). Through the southeastern United States west to East Texas and disjunct to a southwestern limit in the Ottine Wetlands. Reported by Bogusch (1928) and Parks (1935a) and supported by Turner et al. (2003a).

Sawtooth sunflower (*H. grosseserratus* M. Martens), also reported by Parks (1935a) and supported by Turner et al. (2003a), is another moisture-loving sunflower of wooded stream bottoms. It is a large, rhizomatous perennial herb to over 4.5 m tall with glabrous, gray-green, mostly simple stems and shiny, dark green, lance-shaped to egg-shaped-oblong, long-tipped, stalked leaves with coarsely saw-toothed margins. Flower heads small with yellow marginal rays 25–40 mm long and yellow disks 1–2 cm across that flowers during (rarely summer–)October. Throughout the eastern half of the United States and southern Canada to north-central Texas; rare in Southeast Texas, the Ottine Wetlands, and the southern Edwards Plateau (Turner et al. 2003a).

Remarks: From a distance these tall rosinweeds with their broad yellow flower heads might be confused with common sunflowers (*H. annuus* C. Linnaeus). However, the leaves of the common sunflower are broadly triangular with well-developed stalks. They are neither sessile nor elongate. They are also not specific to wetlands and will grow in mesic calcareous as well as sandy soils.

Distribution: Missouri, Oklahoma, south and west through East and Southeast Texas to the West Cross Timbers and eastern Edwards Plateau.

Awnless bush-sunflower
Simsia calva (G. Engelmann & A. Gray) A. Gray
ASTERACEAE (DAISY FAMILY); HELIANTHEAE (SUNFLOWER TRIBE)
Fig. 5-26
Flowering May–November

Field recognition: Low, bushy, sprawling perennial herb from woody roots. Roughly hairy with opposite, dark green, triangular, toothed leaves that are sometimes lobed. Leaves with toothed, leafy, earlike structures (stipules) at the base of the leafstalks. Flower heads single at ends of long stalks with orange-yellow ray and disk flowers. Achenes of the fertile disk florets suggest small, dark, flattened sunflower seeds of the familiar edible kind. Ray florets bearing no seed.

Similar species: None.

Remarks: This is a yellow-flowered relative of the more familiar common sun-flower. It may be seen in uplands adjacent to the wetlands and differs further

Fig. 5-26
Awnless bush-sunflower (*Simsia calva*).

from many of the moisture-loving species here in its preference for calcium-rich soils with a circumneutral pH over acidic sandy or peaty soils. We saw it growing at the pasture margin bordering the North Soefje seepage below the hilltop home of the late Harvey Soefje. Pastures in this area occupy the upper floodplain terrace of the nearby San Marcos River and are regularly blanketed by calcareous, silty alluvium during floods.

Distribution: Occurs widely throughout Texas and northern Mexico except in the eastern and northeastern portions of the state.

Frostweed
Verbesina virginica C. Linnaeus
ASTERACEAE (DAISY FAMILY); HELIANTHEAE (SUNFLOWER TRIBE)
Fig. 5-27
Flowering August–October

Field recognition: Coarse, hairy, green perennial herb to over 2 m with winged stems (continuous with those of the leafstalks), and soft-hairy, alternate, elliptic-lance-shaped, toothed leaves with winged leafstalks. Flower heads with marginal ray florets, borne in densely branched, flat-topped inflorescences at the tops of the plants. Central disk florets "peppered" black and white, and marginal ray florets with white rays to 10 mm long. Fruits flattened, single-seeded achenes with two diverging bristles at the top. Of loamy soils in lowland woods and disturbed areas.

Similar species: None.

Remarks: Frostweed grows abundantly in these wetlands. Its flowers are white, and its leaves provide food for many insects. The broad green platforms are used by male semaphore flies (*Taeniaptera trivittata*) when displaying for females by waving their front legs like flagmen directing a plane onto a landing strip. Frostweed stems are notable for their longitudinal ridges or "wings."

Distribution: The southeastern United States west through East and Southeast Texas to north-central Texas (West Cross Timbers) and the Edwards Plateau.

Fig. 5-27
Frostweed (*Verbesina virginica*).

Cocklebur
Xanthium strumarium C. Linnaeus
ASTERACEAE (DAISY FAMILY); HELIANTHEAE (SUNFLOWER TRIBE)
Fig. 5-28
Fruiting July–November

Field recognition: Coarse, taprooted annual weed of moist, disturbed areas and sandy, gravelly beds of intermittent creeks and streams. Leaves rough-sandpapery, egg-shaped to triangular, coarsely toothed, and often shallowly lobed. Flowers inconspicuous, in unisexual, sessile heads in the axils (angles) of the leaves, the staminate higher and the pistillate lower on the stems. Pistillate heads burlike with two flowers, developing into a reverse egg-shaped, prickly bur covered with hooked prickles and terminating in a pair of vicious-looking spines, to 3 cm long by 1.5 cm wide.

Similar species: None.

Remarks: The photograph shows the better-known fruits rather than the inconspicuous, greenish flowers of cocklebur. These plants can poison pigs if the fruits are eaten, but from the cocklebur's perspective the fruits should be dispersed on the outside of mammals, perhaps in their fur, rather than from the inside via passage through a gut. The prickly fruits are known as "porcupine eggs" and are said to be the inspiration for the invention of Velcro (Diggs, Lipscomb, and O'Kennon 1999).

Distribution: Cocklebur is nearly global within the temperate zone.

Fig. 5-28
Cocklebur (*Xanthium strumarium*) fruits.

Purple marsh-fleabane
Pluchea purpurascens (O. Swartz) A. P. de Candolle
ASTERACEAE (DAISY FAMILY); INULEAE (EVERLASTING TRIBE)
Fig. 5-29
Flowering summer–fall

Field recognition: Aromatic annual herb to nearly 1.5 m, glandular-hairy throughout with egg-shaped to elliptic or lance-shaped, smooth-margined or saw-toothed, sessile or stalked leaves. Flower heads rayless with rose-purple disk florets, and aggregated into flat-topped or layered inflorescences. Fruits tiny, bristle-crowned, single-seeded achenes. Of low, muddy drainage areas.

Similar species: Camphor weed (*P. camphorata* (C. Linnaeus) A. P. de Candolle), a relative of similar habitats in the southeastern United States and the eastern third of Texas west to the Edwards Plateau, may be distinguished by carefully noting that the outer scales (phyllaries) surrounding its flower heads are minutely hairy and fringed along their margins with stiff hairs, whereas the inner ones are more or less smooth except for the tiny golden-yellow, translucent resin globules. In purple marsh-fleabane these inner series of phyllaries are also puberulent and ciliate-fringed. Also the inflorescences of the former tend to be more round-topped than flat-topped or flat-layered.

Remarks: Purple marsh-fleabane grows in muddy places whether in true wetland or along the margin of artificial ponds in upland forest.

Distribution: These purple flowers may be seen in suitable habitats from coast to coast across the southern half of the United States and south through the West Indies and northern South America.

Fig. 5-29
Purple marsh-fleabane (*Pluchea purpurascens*).

Florida lettuce
Lactuca floridana (C.
Linnaeus) J. Gaertner
ASTERACEAE (DAISY FAMILY); LACTUCEAE
(LETTUCE TRIBE)
Fig. 5-30
Flowering August–November

Field recognition: Tall annual herb
with glaucous, blue-green, deeply
feather-lobed, and toothed leaves
with a large triangular tip and with
sticky, milky sap. Flower heads
ligulate; composed entirely of ray
florets with blue or white, tongue-
or strap-shaped ray petals (ligules),
and borne in large, open-branching
inflorescences at the top of the
plant. Fruit a plumed, flattened,
single-seeded achene, slightly
curved and with a very short, stout
beak. Of lowland woods along
streams, creeks, and periodically
flooded areas and low disturbed areas.

Fig. 5-30
Florida lettuce (*Lactuca floridana*).

Similar species: None.

Remarks: "Woodland lettuce," as this species is also known, may exceed 3 m in
height. We found its blue-and-white flowers alongside frostweed in moist soil
like that of the lagoonal area in Palmetto State Park. When the swamp katydid (*Amblycorypha oblongifolia*) matures in early summer, it defoliates this
huge dandelion relative in hordes of fifty or more on a single plant. According to one source (Correll and Johnston 1970) the species is rare in Southeast
Texas, but in the Ottine Wetlands we discovered that it is abundant.

Distribution: Through the eastern United States west through East and Southeast Texas to the West Cross Timbers and Edwards Plateau.

Many-stem false dandelion
Pyrrhopappus pauciflorus (D. Don) A. P. de Candolle
ASTERACEAE (DAISY FAMILY);
LACTUCEAE (LETTUCE OR DANDELION TRIBE)
Fig. 5-31
Flowering March–June

Field recognition: Annual herb from a tapering taproot with milky sap and
stems few-branched from a basal rosette of feather-lobed or toothed leaves,
stem leaves reduced upward. Flower heads composed entirely of ray florets

Fig. 5-31
Many-stem
false dandelion
(*Pyrrhopappus*
pauciflorus).

(ligulate), each with a tongue- or strap-shaped petal (ligule). Solitary, large, and showy, yellow and open during the morning. Fruits beaked, single-seeded achenes crowned with a ring of hairlike bristles. Preferring clay soils of prairies and roadsides.

Similar species: Carolina false dandelion (*P. carolinianus* (T. Walter) A. P. de Candolle) is a similar annual species of the southeastern United States through East and Southeast Texas to the West Cross Timbers that occurs mostly in sandy soils of woodlands and fields. It is along the southern margin of its range here. This relative may be distinguished by its mostly unbranched, smooth, hairless stems and the outer (basal) scales of its flower heads that are one- to two-thirds as long as the inner rather than a third or less (flowering April–July). These two annual species hybridize where human disturbance has brought sandy and clayey soils into close proximity, such as along highways where sand and gravel have been brought in for fill (Diggs, Lipscomb, and O'Kennon 1999). Tuber false dandelion (*P. grandiflorus* (T. Nuttall) T. Nuttall) (April–May), a prairie species of the southern Great Plains, is a rhizomatous perennial from a tuber or tuberous, thickened roots.

Remarks: Many-stem false dandelion, a white- and yellow-flowered "weed," was encountered on the floodplain of the San Marcos River, along a trail inside Palmetto State Park, despite the fact that this is not a wetland plant in the strict sense.

Distribution: The species ranges from northern Mexico through South, Central, Southeast, and north-central Texas to Oklahoma.

Leafy elephant's foot
Elephantopus carolinianus E. Raeuschel
ASTERACEAE (DAISY FAMILY); VERNONIEAE (IRONWEED TRIBE)
Fig. 5-32
Flowering August–October

Field recognition: Pubescent perennial herb to 1 m tall with reverse lance-shaped-elliptic, alternate leaves. Basal leaves withered or absent by flowering time; stem leaves conspicuous, larger. Flower heads secondary, compounded of smaller, primary heads of 3–4 florets aggregated into dense clusters (glomerules) at the ends of upper branches and surrounded by three leaflike, triangular bracts (about 1 cm long). Florets with tubular pinkish-white to lavender petals. Fruits cylindric, single-seeded achenes with a crown of 5–10 stiff awns. Of low, sandy, wooded areas, usually moist and shaded, such as mesic slopes, floodplains, and seepage areas.

Similar species: None.

Remarks: The pattern of secondary aggregation of numerous reduced, primary flower heads is seen in several other unrelated genera of the Asteraceae and in members of other unrelated families as well, such as the poinsettia (*Euphorbia pulcherrima* Willard *ex* Klotzsch, Euphorbiaceae). The pattern seems to belie an evolutionary history of floral reduction, followed by its re-evolution via multiplication and aggregation of the reduced flowers (florets) and elaboration of neighboring leaves. The result is a structure that appears to functionally simulate or mimic a true flower, but it is actually a secondary structure composed of many reduced flowers and nearby nonfloral structures, like the red, petal-like leaves surrounding the cyathia of a poinsettia. Oft-repeated patterns such as these seem testaments to the oscillating nature and imperfect reversibility of evolution.

Distribution: The southeastern United States through East Texas to north-central (Grand Prairie) and Southeast Texas to its southwestern margin or limit in the Ottine Wetlands.

Fig. 5-32
Leafy elephant's
foot (*Elephantopus
carolinianus*).

Fig. 5-33
Turnsole (*Heliotropium indicum*).

Turnsole
Heliotropium indicum C. Linnaeus
BORAGINACEAE (FORGET-ME-NOT FAMILY)
Fig. 5-33
Flowering June–October

Field recognition: Coarse, hairy annual herb to over 1 m tall. Growing in moist disturbed areas and along shores, banks, and river bottoms. Leaves large, opposite, stalked, egg-shaped to elliptic. Flowers blue to violet, borne in two rows along the upper side of a coiled, elongate cluster (a helicoid or scorpioid cyme) that unrolls as they open.

Similar species: None.

Remarks: Turnsole is an exotic poisonous plant with lavender flowers that has been variously identified as native to the Old World, to the New World, and to Texas, with disagreement about its precise natural distribution (Diggs, Lipscomb, and O'Kennon 1999). We first identified the plant in the lagoon region of Palmetto State Park.

Distribution: Throughout warmer parts of the New World, probably native, from northern Argentina to the southern United States through East and Southeast Texas (Correll and Johnston 1970).

Spring forget-me-not
Myosotis macrosperma G. Engelmann
BORAGINACEAE (FORGET-ME-NOT FAMILY)
Fig. 5-34
Flowering (March–)April–early May

Field recognition: Small, hairy, erect annual herb with spoon-shaped to reverse lance-shaped leaves, and tiny white or slightly bluish flowers borne on greatly elongating, loosely flowered clusters arising from axils of the upper leaves.

Fruit four smooth, ovoid seeds (mericarps) enclosed by the mature sepals and covered with hooked hairs. Of rich sandy or silty soils of woods, bottomlands, stream banks, and roadsides.

Similar species: None.

Remarks: Its small burlike fruits covered with Velcro-like hooked hairs may be adapted for dispersal in the fur of small mammals.

Distribution: Through the southeastern United States west through East and Southeast Texas to the West Cross Timbers and Edwards Plateau. Spring forget-me-not is near the southwestern limit of its range here.

Fig. 5-34
Spring forget-me-not (*Myosotis macrosperma*).

Bejar marbleseed

Onosmodium bejariense A. P. de Candolle *ex* A. L. de Candolle subsp. *bejariense* var. *bejariense*

BORAGINACEAE (FORGET-ME-NOT FAMILY)

Fig. 5-35

Flowering April–July

Field recognition: Erect or ascending white-hairy perennial herb from a woody crowned taproot, with lance-shaped, reverse lance-shaped to elliptic leaves conspicuously veined and glossy green beneath the covering of long, white hairs. Flowers borne in terminal or nearly terminal branched, leafy-bracted, 1-sided clusters (scorpioid cymes) that uncurl as the flowers successively open. Flowers tubular, greenish to yellowish-white, appearing swollen below the five erect petal lobes surrounding the protruding styles that bear the pollen-receiving stigmas. Fruit up to four white to dingy brownish-white, smooth spherical seeds. Of grasslands and forest openings in deep silty to silty-clay soils (Turner 1995) and limestone outcrops (Diggs, Lipscomb, and O'Kennon 1999).

Similar species: None.

Remarks: Marbleseed's yellowish-white flowers appear to be closed, but they are open, the styles protruding beyond the surrounding petals. Pollinating insects such as bumblebees manually open them to gain access to pollen and nectar. We found this plant in the wetlands floodplain despite the fact that it is reported as a species of dry woodlands and hillsides (Correll and Johnston 1970).

Fig. 5-35
Bejar marbleseed
(*Onosmodium
bejariense*).

Distribution: Bejar marbleseed var. *bejariense* has been reported as a Texas endemic, but the subspecies, including var. *hispidissimum* (K. Mackenzie) B. L. Turner and var. *occidentale* (K. Mackenzie) B. L. Turner, occurs throughout the central United States east of the Rocky Mountains and the interior eastern United States from southern Canada south through East and Southeast Texas to the West Cross Timbers and southern Edwards Plateau (Turner 1995; Diggs, Lipscomb, and O'Kennon 1999).

Cardinal flower
Lobelia cardinalis C. Linnaeus
CAMPANULACEAE (BLUEBELL FAMILY);
LOBELIOIDEAE (LOBELIA SUBFAMILY)
Fig. 5-36
*Flowering (May–)September–
October(–December)*

Field recognition: Perennial herb via basal offshoots to over 2 m tall with erect, simple stems and egg-shaped to lance-shaped, irregularly finely saw-toothed leaves. Flowers showy, borne in elongate terminal spikes, bright, deep crimson or vermilion-red, 2-lipped, long-tubular (3–5 cm long) with tube (1.5–2.0 cm long) and with slitlike openings near the base (fenestrate). Pollen-bearing stamens also united into a tube through which the style is exserted, expelling pollen. Fruit capsule

Fig. 5-36
Cardinal flower (*Lobelia cardinalis*).

contains numerous seeds. Of low, moist, or wet open areas along streams, springs, ponds, swamps, and moist meadows and along roadside ditches.

Similar species: None.

Remarks: The cardinal flower's bright red colors may be seen in Rutledge Swamp and in North Soefje Swamp, where we stood nearby while a female ruby-throated hummingbird (*Archilochus colubris*) flew in and sipped nectar within arm's reach, dusting the top and back of its head with pollen in the process. This flower appears to be ideally adapted for pollination by migrating ruby-throats and other hummingbird species farther west. However, the situation may be a bit more complex, as the corolla tube is fenestrate, which would seem to permit nectar theft by bees, and some populations may produce little or no nectar at all. Its outward beauty belies its potentially lethal chemistry, for the entire plant is poisonous.

Distribution: Throughout the United States; southern Canada to Mexico.

Ponyfoot
Dichondra carolinensis A. Michaux
CONVOLVULACEAE (MORNING GLORY FAMILY)
Fig. 5-37
Flowering (March–)April(–June)

Fig. 5-37
Ponyfoot (*Dichondra carolinensis*).

Field recognition: Creeping, mat-forming perennial herb from horizontal stems rooting at the nodes (points of leaf attachment) in damp silty or sandy ground. This plant is also a weed of lawns. Leaves small, rounded, on long stalks. Flowers very small, arising from the axils (angles) of the leafstalks, with yellow-green petals and borne at the ends of erect stalks.

Similar species: *Dichondra micrantha* I. Urban is a species of similar habitats in South Texas and adjacent Mexico that flowers from April to May. It has flowers with white petals on flower stalks that are recurved near their summits so that the flowers face downward.

Remarks: The leaves shown in the photograph that resemble the hoofprint of a horse are those of the aptly named ponyfoot, a plant of damp soil in the wild but also a weed on watered lawns.

Distribution: Ponyfoot occurs from throughout the southeastern quarter of the United States to the West Cross Timbers of Texas.

Snow-on-the-mountain

Euphorbia (Agaloma A. Love & D. Love) *marginata* F. Pursh
EUPHORBIACEAE (SPURGE FAMILY)
Fig. 5-38
Flowering (late spring–)July–October

Field recognition: Stout annual herb with caustic milky sap and elliptic, oblong or egg-shaped, smooth, green lower stem leaves and smaller, narrowly lance- to egg-shaped, white-margined, leaflike bracts near the top of the plant surrounding the inflorescences. Inflorescences (cyathia) cup- or bell-shaped with five conspicuous, cupped green glands on the margin with white, petal-like appendages. Cyathia containing many pollen-bearing, staminate flowers (reduced to single stamens) and one pistillate flower with its ovary exserted and bent to one side on a stalk (gynophore), globose and 3-lobed. Fruit a flattened-globose, 3-celled capsule containing seeds. Of calcareous uplands, stream bottoms, and low areas.

Similar species: Snow-on-the-prairie (*E.* (*A.*) *bicolor* G. Engelmann & A. Gray) of the eastern third of Texas in tight, clay soils of blackland and grand prairies eastward, a related annual herb, differs in having narrow, lance-shaped to narrowly elliptic stem leaves and linear, white-margined bracts. However, intermediate individuals occur where the ranges of two species overlap, indicating possible introgressive hybridization (Diggs, Lipscomb, and O'Kennon 1999).

Remarks: Like the related poinsettia (*E. pulcherrima* C. Willdenow *ex* J. Klotzsch), snow-on-the-mountain is white by virtue of its upper leaves (red in poinsettia) rather than its flowers (aggregated in cuplike structures called cyathia). It is characterized as an upland plant, but we found it along the interface between North Soefje Swamp and the adjacent pasturelands. The plant produces an "evil-tasting poisonous honey" toxic to livestock, its sap causes skin lesions, and it has been used to brand cattle in lieu of a hot iron (Diggs, Lipscomb, and O'Kennon 1999).

Fig. 5-38
Snow-on-the-mountain (*Euphorbia marginata*).

Distribution: Unusually western for a species of the Ottine Wetlands, for it occurs in the central Great Plains states from Montana to Minnesota and south to New Mexico through the western two-thirds of Texas and is adventive eastward along highways to the Gulf Coast (Turner et al. 2003a).

Mat euphorbia
Euphorbia (Chamaesyce J. K. Small) *serpens* K. Kunth
EUPHORBIACEAE (SPURGE FAMILY)
Fig. 5-39
Flowering (March–)July–October(–November)

Fig. 5-39
Mat euphorbia (*Euphorbia serpens*).

Field recognition: Prostrate, smooth, mat-forming annual (possibly over-wintering as winter annual) herb with milky sap, often rooting at the nodes. Stems slender and many-branched with opposite, rounded-egg-shaped to broadly oblong, entire, short-stalked leaves. Cuplike inflorescences (cyathia) solitary in leaf angles and at the nodes. Gland appendages along the margins of the inflorescence cups irregularly toothed, seeds smooth. In calcareous clay soils of stream bottoms, clay flats, and prairies or as a weed in disturbed areas.

Similar species: Eyebane (*E. (C.) nutans* M. Lagasca y Segura), a widespread species of similar habitats throughout Texas and the eastern United States and tropical/temperate regions worldwide, is an erect annual with saw-toothed, often red or maroon-blotched leaves and pubescent stems (May–October[–November]). Ridge-seed euphorbia (*E. (C.) glyptosperma* G. Engelmann), another widespread North American species of similar habitats but in sandy soils, is very similar but has seeds with transverse ridges rather than smooth surfaces ([late May–]June–September[–October]). It occurs throughout Texas except in the eastern portion of the state (Turner et al. 2003a).

Remarks: This plant has tiny, white, flowerlike structures (pseudanthia: "false-flowers") called cyathia that are actually cuplike structures containing several reduced male flowers (stamens) and one reduced female flower (a stalked ovary).

Distribution: Widespread within the tropical and temperate New World and throughout Texas.

Coffee senna
Senna occidentalis (C. Linnaeus) J. Link
FABACEAE (BEAN FAMILY); CAESALPINIODEAE
(CASSIA SUBFAMILY)
Fig. 5-40
Flowering August–November

Field recognition: Annual, ill-
smelling herb to over 2 m tall with
alternate, even-feather-compound
(lacking a terminal leaflet) leaves of
4–6 pairs of lance-shaped to egg-
shaped leaflets, the distal pairs the
largest. Flowers yellow or yellow-
orange, with five undifferentiated
petals in few-flowered, spikelike
clusters in the leaf angles (axils).
Fruits erect pods, dark brown with
lighter margins when mature.

Similar species: None.

Remarks: Coffee senna is a weed that
grows along the edge of North
Soefje Marsh, though it is by no
means restricted to wet habitats.
Native acrobatic ants (*Crematogaster* sp.) and exotic yellow meal-

Fig. 5-40
Coffee senna (*Senna occidentalis*) with acrobatic
ants (*Crematogaster* sp.) and mealworm beetles
(*Tenebrio molitor*) feeding at its (nectar?) glands.

worm beetles (*Tenebrio molitor*) feed at glands (extrafloral nectaries?) on the
leafstalks.

Distribution: This exotic plant probably originated in the American Tropics. It
is a pantropical weed of old pastures north through Southeast and East Texas
and inland to the Edwards Plateau and the southeastern coastal United States.

Sandyland Texas bluebonnet
Lupinus subcarnosus W. Hooker
FABACEAE (BEAN FAMILY); PAPILIONOIDEAE (PEA SUBFAMILY)
Fig. 5-41
Flowering March–April

Field recognition: Winter annual herb from a rosette of fan-lobed leaves on
long stalks, sprawling branches appearing later. Leaflets usually five in number (rarely as many as seven), reverse egg-shaped to reverse lance-shaped.
Flowers borne severally on narrow, rounded, open spikes. Flowers bright blue
to purplish or violet with inflated, lateral wing petals. Upper banner petal
with a white basal spot. Endemic in sandy to loamy soils of the Post Oak
Savannah region of Texas.

Similar species: Texas bluebonnet (*L. texensis* W. Hooker) is another lupine endemic in calcareous, clayey soils of the Edwards Plateau and Blackland Prairie regions of the state. Originally segregated from its sandyland congener by habitat and soil preference, it has now spread widely throughout the state along highway margins and in overgrazed pastures, as well as via cultivation. It may be distinguished by its thicker, pointed, more crowded spikes of dark blue flowers with flat, uninflated lateral wing petals that appear straight rather than rounded in frontal view.

Remarks: Six bluebonnet species compose the set of "officially legislated" Texas state flowers. The one shown in the photograph grows in sandy soils, whereas *L. texensis* requires limestone-type soils. Neither is a wetland plant, though both may be seen in the Ottine area.

Fig. 5-41
Sandyland Texas bluebonnet (*Lupinus subcarnosus*).

Distribution: Our featured species is Texas endemic as well as Texas legal, occurring only in sandy to loamy soil corridors of the Post Oak Savannah in the southeast-central part of the state.

Texas-bluebell, showy prairie-gentian
Eustoma russellianum (W. Hooker) G. Don
(*E. grandiflorum* (C. Rafinesque-Schmaltz) L. Shinners)
GENTIANACEAE (GENTIAN FAMILY)
Fig. 5-42
Flowering late June–August(–September)

Field recognition: Erect, glaucous-green, taprooted annual or short-lived perennial herb with opposite, 3-veined egg-shaped to elliptic-oblong or elliptic-lance-shaped leaves, to nearly 1 m high. Flowers large, showy, deeply bell-shaped or tuliplike, blue-purple with darker or sometimes yellow center. Bright yellow, flattened, 2-lobed, pollen-receiving stigma prominent in the flower center. Fruit a many-seeded capsule to 2 cm long. Of moist prairie swales, fields, and other low, moist, open areas and sometimes disturbed areas around tanks.

Similar species: The smaller-flowered *E. exaltatum* (C. Linnaeus) A. Salisbury *ex* G. Don has been distinguished on the basis of flower size (petal lobes ≤25 mm long by <15 mm wide, versus >30 mm long by ≥20 mm wide). However, considerable intergradation exists (Diggs, Lipscomb, and O'Kennon 1999), and the two forms may not be different species. In that case *E. russellianum* would properly be known as *E. exaltatum* var. *russellianum.*

Remarks: One of our most striking and beautiful wildflowers.

Distribution: Throughout the south-central United States from Oklahoma, Nebraska, and Colorado south through Texas to Mexico.

Fig. 5-42
Texas-bluebell, showy prairie-gentian (*Eustoma russellianum*).

Prairie rose-gentian, Texas star
Sabatia campestris T. Nuttall
GENTIANACEAE (GENTIAN FAMILY)
Fig. 5-43
Flowering (April–)May–early July

Field recognition: Erect annual herb to half a meter in height, divaricately branching above with opposite, stalkless, oblong-elliptic to broadly egg-shaped-elliptic leaves clasping and surrounding stem at their bases. Flowers showy, five petals, rose or pale pink, each with a triangular yellow or greenish

Fig. 5-43
Prairie rose-gentian (*Sabatia campestris*).

spot at base bordered with a white and/or reddish area, together appearing as a yellow star in the flower center. Fruit a dry, many-seeded capsule, splitting open at maturity. In moist or dry, usually clay soils of prairies, fields, wood edges, and roadsides and along streams.

Similar species: None. (But compare Maryland meadow-beauty [*Rhexia mariana*].)

Remarks: Texas star blooms with a pinkish-red flower and is treated here because of its beauty and name and proximity to wetlands, for we saw it as we moved between marshes and swamps rather than when we were passing through them.

Distribution: From Illinois to Kansas and south from Mississippi through the eastern half of Texas (Correll and Johnston 1970).

Baby blue eyes
Nemophila phacelioides T. Nuttall
HYDROPHYLLACEAE (WATERLEAF FAMILY)
Fig. 5-44
Flowering (March–)April–May

Field recognition: Annual herb of sandy or silty stream bottom woods and moist lowland openings, with oblong feather-lobed or feather-compound leaves composed of egg-shaped divisions or leaflets. Flowers showy, wheel-shaped, 5-petaled, blue-lavender with pale central "eye," solitary from the leaf axils or in few-flowered terminal clusters. Fruit a globose capsule and enclosed by the pointed sepal lobes.

Similar species: None.

Remarks: Baby blue eyes with its lavender flowers is tied more strongly to sand than to wet soil.

Distribution: This plant occurs in a relatively restricted range from Arkansas and Oklahoma south and west through East and Southeast to north-central Texas and the Edwards Plateau.

Fig. 5-44
Baby blue eyes
(*Nemophila phacelioides*).

Lemon beebalm
Monarda citriodora V. de Cervantes
LAMIACEAE (MINT FAMILY)
Fig. 5-45
Flowering (April–)May–July(–October)

Field recognition: Aromatic, finely hairy annual or perennial herb with lance-shaped to oblong-lance-shaped, toothed leaves and 2-lipped, tubular white to pink or lavender flowers with purple spots in their throats. Flowers arranged in dense, whorled heads in interrupted terminal spikes, each headlike whorl surrounded by spreading, reflexed leaves (bracts) with purple or lavender-pink, pubescent inner surfaces. Bracts with bristle-like tips. Plant containing citronellol. Of mostly upland sandy or rocky soils in prairies, in meadows, and along roadsides.

Similar species: Spotted beebalm (*M. punctata* C. Linnaeus) (Fig. 5-46), another sandy upland species that may be observed abutting these wetlands, may be distinguished by noting that the bracts surrounding the headlike whorls of flowers do not end in elongate hair- or bristle-like tips but are simply short- or long-pointed.

    Wild bergamot (*M. fistulosa* C. Linnaeus) was reported for these wetlands by Bogusch (1928) and Parks (1935a). It is a species of open woods, edges, old fields, wet meadows, marshes, ditches, and stream banks in the eastern United States that reaches its southwestern limit in north-central and East Texas. This record constitutes a southwestern disjunction for the species if accurate. It is distinctive in hav-

Fig. 5-45
Lemon beebalm (*Monarda citriodora*).

Fig. 5-46
Spotted beebalm (*Monarda punctata*).

ing a single flower head at the ends of the branches. Flowers lavender or white with a flared tube and 2-lipped with stamens exserted beyond the upper lip. However, Lindheimer's beebalm (*M. lindheimeri* G. Engelmann & A. Gray *ex* A. Gray), with shorter-stalked midstem leaves (3–7 mm long versus ≥8 mm long) with long, spreading hairs beneath, and white flowers, is supported by a collection record from Gonzales County (Turner et al. 2003a). It ranges through the eastern quarter of Texas and Louisiana with a southwestern disjunction here. The *M. fistulosa* of Bogusch and Parks may refer to this species.

Remarks: The white and purple-flecked flowers of lemon beebalm appear in several tiers and attract a variety of insects that seek its pollen and nectar. The lemon scent for which the species is named arises from the leaves and stems, which contain citronellol. This aromatic compound is used as an insect repellent by humans and may serve the plant in a similar capacity. This is actually an upland plant that may be seen while moving to and from the swamps and marshes.

Distribution: Lemon beebalm occurs throughout the southern Great Plains, south throughout Texas to northeastern Mexico, and is adventive eastward to near the Mississippi River.

Mealy blue sage
Salvia farinacea G. Bentham
LAMIACEAE (MINT FAMILY)
Fig. 5-47
Flowering April–July

Field recognition: Perennial herb from a woody root, minutely powdery-pubescent with lance-shaped to oblong or lance-egg-shaped, stalked, toothed leaves with tapering bases. Flowers borne in whorls along interrupted terminal spikes with 2-lipped purple or violet-blue tubular petals, and felty white or purplish matted hairs on the united sepals. In calcareous, limestone-derived soils of floodplain meadows, prairies, thickets, and upland habitats throughout most of Texas. Often cultivated as an ornamental.

Fig. 5-47
Mealy blue sage (*Salvia farinacea*).

Similar species: None.

Remarks: The blue and white petals of mealy blue sage appear not only in flood-plains, where we found them, but also in uplands. It prefers calcareous soils brought in by the San Marcos River as opposed to sandy soil derived from the wetland's own underlying Carrizo Formation.

Distribution: Mealy blue sage is native to Central and West Texas, New Mexico, and northern Mexico and is scattered eastward.

Lax hornpod
Mitreola petiolata (T. Walter) J. Torrey & A. Gray (*Cynoctonum mitreola* (C. Linnaeus) N. Britton)
LOGANIACEAE (STRYCHNINE FAMILY)
Fig. 5-48
Flowering (May–)June–October(–November)

Fig. 5-48
Lax hornpod (*Mitreola petiolata*).

Field recognition: Smooth, hairless annual herb with simple to few-branched stems with opposite, thin, light green, short-stalked egg-shaped-elliptic to elliptic-lance-shaped leaves with small, leafy appendages (stipules) near their bases. Flowers small, white, somewhat urn-shaped and borne in dense, 1-sided, long-stalked, spike-like clusters arising in opposite pairs from the axils of the upper leaves. Fruits very distinctive 2-horned capsules or pods. Of wet sandy or peaty soils and muck of seepage areas, ditches, and streams and near ponds and lakes.

Similar species: None.

Remarks: Unlike most of the other species of this family treated in the standard reference to plants of Central Texas, lax hornpod is not characterized as a poisonous plant (Diggs, Lipscomb, and O'Kennon 1999). The lethal poison and medicine known as "strychnine" is obtained from an Old World species belonging to the same family.

Distribution: The small white flowers of lax hornpod appear in wetlands throughout the southeastern quarter of the United States west through East and Southeast Texas to the southern Edwards Plateau.

Florida pinkroot (Texas pinkroot in part)

Spigelia loganioides (J. Torrey & A. Gray *ex*
S. F. Endlicher & E. Fenzl) A. L. de Candolle
(*S. texana* (J. Torrey & A. Gray) A. L. de Candolle in part)
LOGANIACEAE (STRYCHNINE FAMILY)
Fig. 5-49
Flowering (spring–)June–August

Field recognition: Low, smooth, spreading perennial herb with opposite, egg-shaped to elliptic-lance-shaped or reverse egg-shaped to reverse lance-shaped leaves, with reduced, leaflike appendages at their bases (stipules). Flowers terminal and in opposite leaf axils above a whorl of four stem leaves. Corolla funnel-shaped, white with faintly pink veins. Fruit a very distinctive 2-celled twin pod, somewhat heart-shaped in outline. Of rich, mesic wooded slopes, riparian forests, floodplain woods, and stream banks in black-clay, limestone, or chalk-derived soils.

Similar species: Prairie pinkroot (*S. hedyotidea*), a sister species of drier, upland limestone-derived soils, is known immediately to the south and west of the Ottine Wetlands, through Central Texas to central Mexico. It may be distinguished by its shorter, bushier habit; smaller, thicker, narrower leaves; and smaller flowers (Henrickson 1996).

Remarks: Florida pinkroot is a white-flowered poisonous species belonging to the family of plants that produce strychnine.

Distribution: Texas pinkroot (*S. texana*), formerly regarded as a Southeast Texas endemic, has recently been interpreted as conspecific with Florida pinkroot (*S. loganioides*) (Henrickson 1996). It is known only from Texas and Florida but not the intervening states. Pinkroot prefers calcareous soils over the sandy, Coastal Plain soils such as those derived from the Carrizo Formation that makes these swamps and marshes possible, and this may in part explain its current disjunct distribution (Henrickson 1996).

Fig. 5-49
Florida pinkroot
(*Spigelia loganioides*).

Wright's false mallow
Malvastrum aurantiacum
(G. Scheele) W. Walpers
MALVACEAE (MALLOW FAMILY)
Fig. 5-50
Flowering April–July(–October)

Field recognition: Low, shrubby perennial from a woody base with broadly egg-shaped to oblong leaves coarsely scalloped-toothed and grayish to dull green due to a covering of scurfy, star-shaped hairs. Flowers borne in the leaf axils (angles), five petals, asymmetrically diamond-shaped and golden to pale orange-yellow, pinwheel-shaped and hibiscus-like with a central stamen column or tube, bearing pollen-shedding anthers on the outside. Fruits brownish-red when mature, of flattened, disklike sections arranged radially in center of the sepals and surrounding leaves.

Fig. 5-50
Wright's false mallow (*Malvastrum aurantiacum*).

Similar species: Three-lobe false mallow (*M. coromandelianum* (C. Linnaeus) G. Garcke) (Fig. 5-51), an herbaceous annual or perennial pantropical weed of Central and South Texas, is similar but has smaller flowers and sharply toothed egg-shaped to lance-shaped leaves that appear smooth and dark green above due to the lack of dense, scurfy, star-shaped hairs (it has only sparse, simple, or star-shaped pubescence). It is a plant of clayey, alkaline soils and is often weedy in disturbed areas, flowering virtually year-round (espe-

Fig. 5-51
Three-lobe false mallow (*Malvastrum coromandelianum*).

cially April–November). It may be abundantly seen in nearby upland areas adjoining these wetlands.

Remarks: Wright's false mallow is a yellow-petaled Texas endemic that grows along watercourses and in periodically flooded pastures in the Ottine area (Diggs, Lipscomb, and O'Kennon 1999).

Distribution: An older source reports the species as rare, limited to Central and South Texas, but likely to exist in northern Mexico (Correll and Johnston 1970). The presumed endemic was only recently reported from the Ottine region (Turner et al. 2003a).

Turk's cap, Texas mallow
Malvaviscus arboreus J. Dillenius
ex A. Canavilles var. *drummondii*
(J. Torrey & A. Gray) R. Schery
MALVACEAE (MALLOW FAMILY)
Fig. 5-52
Flowering June–July(–year-round)

Field recognition: Many-branched shrub to 3 m high with finely pubescent, alternate, stalked, rounded-heart-shaped leaves shallowly 3-lobed to angled and broadly, bluntly toothed. Flowers borne in the upper leaf axils, showy bright red or scarlet, five petals remaining twisted and tubelike, only slightly spreading about the exserted stamen tube and red, hairy-knobbed stigmas. Fruits 5-celled, spongy-fleshy, red and berrylike with a flavor like watermelon and containing five stones. In limestone-derived soils of slopes, ledges, and arroyos; along streams and at the edges of thickets.

Fig. 5-52
Turk's cap (*Malvaviscus arboreus*).

Similar species: None.

Remarks: Turk's cap, also known as "Texas mallow," is a red-flowered plant commonly encountered in Central Texas, whether in the shade of hackberries in the capital city of Austin, or in the shade of the same trees along the oxbow lake in Palmetto State Park. Its preference is for limey soils rather than those derived from the Carrizo Sands. This is yet another species with flowers that appear ideally adapted for pollination by hummingbirds.

Distribution: The coastal United States from Florida and Cuba through Southeast Texas to the southern Edwards Plateau and southeastward to adjacent Mexico.

Carolina modiola
Modiola caroliniana (C. Linnaeus) G. Don
MALVACEAE (MALLOW FAMILY)
Fig. 5-53
Flowering (March–)April–June

Field recognition: Low, creeping or open-branching perennial herb with rounded egg-shaped, shallowly to deeply, 3–5 fan-lobed and toothed leaves. Flowers small, borne on stalks in the axils (angles) of the leaves. Salmon to purplish-red, *Hibiscus*-like with stamen-tubes. Fruits depressed, wheel-like, of 15–30 thin, disklike sections, each late opening and containing a couple of seeds. Of various more or less moist areas and disturbed habitats such as saltmarsh edges, lawns, and gardens.

Similar species: None. (But compare three-lobe false mallow [*Malvastrum coromandelianum*].)

Remarks: The petals of Carolina modiola appear in various reddish shades such as the salmon color of the flower in the photograph. Its habitats are varied, but the borders of saltmarshes are among them, so they might be expected near the Gulf cordgrass marsh of Palmetto State Park. We found them in open floodplain and disturbed habitats throughout the area.

Distribution: The southeastern United States through East and Southeast Texas to Argentina and widely scattered as a weed elsewhere.

Fig. 5-53
Carolina modiola
(*Modiola caroliniana*).

Maryland meadow beauty
Rhexia mariana C. Linnaeus var. *interior*
(F. Pennell) R. Kral & P. Bostick
MELASTOMATACEAE (MEADOW BEAUTY FAMILY)
Fig. 5-54
Flowering May–September

Field recognition: Erect perennial herb to three-quarters meter tall from elongate rhizomes near the soil surface, with 4-sided stems and opposite, simple linear to elliptic, egg-shaped or lance-shaped leaves with three fan to parallel, arc-shaped (curving along leaf margins) main veins and finely toothed to ciliate (straight-hairy) margins. Flowers showy, in densely branching inflorescences at top of plant. Four petals, reverse egg-shaped and bright rose-pink with eight stamens having conspicuous, bright yellow, elongate, curved anthers opening by terminal pores. Fruits erect, urn-shaped capsules topped by a short tube formed by the fused sepals. Of moist or wet open areas such as wet meadows, seepage areas, bogs, and ditches and at the edges of moist thickets.

Similar species: None.

Remarks: Meadow beauty's pink flowers bloom in bogs and along wetland seepages from the Atlantic Ocean to Texas, though the species is apparently absent from Florida (Correll and Johnston 1970).

Distribution: The southeastern United States to East and Southeast Texas to its southwestern limit in the Ottine Wetlands (Kral and Bostick 1969).

Fig. 5-54
Maryland meadow beauty (*Rhexia mariana*).

Fig. 5-55
Yellow cow-lily
(*Nuphar advena*).

Yellow cow-lily, spatter-dock
Nuphar advena (C. Linnaeus) J. E. Smith
NYMPHEACEAE (WATER-LILY FAMILY)
Fig. 5-55
Flowering (March–)June–October

Field recognition: Perennial, rhizomatous aquatic plant with large, rounded, floating or emergent leaves that are shiny and dark green above, with a deep notch at the attachment point of the leafstalk. Flowers cup- or goblet-shaped with six concave tepals (similar petals and sepals) that are rich yellow inside and green to yellow-green outside. Fruit an ovoid capsule, constricted below the pollen-receiving disk (stigma) and maturing underwater.

Similar species: None.

Remarks: Spatter-dock is not a water-lily in the usual sense of that common name, but it is a close relative. The yellow flowers appear on the surface of a few ponds in these wetlands and the surrounding pastures. Though the blooms are held above water, facilitating pollination by flying insects, the fruit matures beneath the surface.

Distribution: Throughout eastern North America to East Texas and the Edwards Plateau to northern Mexico; also Cuba.

Shrubby water-primrose
Ludwigia octovalvis (N. von Jacquin) P. Raven
ONAGRACEAE (EVENING-PRIMROSE FAMILY)
Fig. 5-56
Flowering July–October

Field recognition: Shrubby, branching perennial herb to over 1 m tall with narrowly lance-shaped to narrowly egg-shaped, alternate leaves. Flowers showy, with four yellow, rounded petals and eight stamens. Fruit an elongate capsule, splitting when mature, revealing many small, dark seeds packed into four rows. Of swamps, pond margins, and other wet places.

Similar species: Bushy seedbox (*L. alternifolia* C. Linnaeus), reported by both Bogusch (1928) and Parks (1935a), is similar but has short seed capsules (2–10 [rarely 12] mm long versus 17–45 mm long) and four rather than eight pollen-bearing stamens per flower ([April–]June–August[–September]). It ranges throughout the southeastern United States through East and Southeast Texas to the East Cross Timbers and Ottine Wetlands.

Remarks: The delicate yellow flower petals are fugacious (ephemeral), opening in the morning and dropping at the slightest touch by early afternoon of the same day.

Fig. 5-56
Shrubby water-primrose (*Ludwigia octovalvis*).

Distribution: In wet areas throughout the warmer regions of the world, including the southern half of Texas.

Large evening-primrose
Oenothera grandis (N. Britton) B. Smyth
(*O. laciniata* J. Hill var. *grandiflora* (S. Watson) B. Robinson)
ONAGRACEAE (EVENING-PRIMROSE FAMILY)
Fig. 5-57
Flowering (March–)April–June

Field recognition: Annual herb with wavy-margined, toothed to feather-lobed leaves. Flowers showy with four deep yellow petals united into a tube basally and opening from late afternoon to near sunset. Of sandy, open, mesic and seasonally moist areas.

Similar species: Cutleaf evening-primrose (*O. laciniata* J. Hill) is vegetatively similar but has much smaller, pale yellow flowers

Fig. 5-57
Large evening-primrose (*Oenothera grandis*).

opening from late afternoon to late morning. It may be less dependent on strictly nocturnal pollinators. The cutleaf species is an annual to short-lived perennial flowering from March to November throughout the eastern United States and Texas south to Ecuador.

Remarks: With showy, tubular flowers opening in late afternoon to near sunset, this species may be reliant on nocturnal and crepuscular moths for cross-pollination.

Distribution: This evening-primrose ranges from the south-central Plains states south through East, Central, and South Texas to northern Mexico.

Showy primrose
Oenothera speciosa T. Nuttall
ONAGRACEAE (EVENING-PRIMROSE FAMILY)
Fig. 5-58
Flowering April–July

Field recognition: Low, rhizomatous perennial herb with wavy-margined to feather-lobed, reverse egg-shaped to oblong-lance-shaped leaves. Flowers borne in short, nodding spikes. Large and conspicuous with short tubes and four large rose-pink to white petals, opening in morning (pink) or evening (white). Fruit tough, club-shaped, sessile capsules with the lower, constricted, cylindrical portion sterile (without seeds) and the widened upper part containing several rows of tiny (1 mm) seeds. Of prairies and open woodlands, the rose-pink forms are cultivated and often escape as weeds in lawns and along roadsides.

Similar species: None.

Remarks: Showy primrose is an abundant and attractive pink or white wildflower, though it is not a wetland plant per se.

Distribution: This evening-primrose ranges from the southern, central plains to the Mississippi River and throughout Texas to northeastern Mexico.

Fig. 5-58
Showy primrose
(*Oenothera speciosa*).

Stemless evening-primrose
Oenothera triloba T. Nuttall
ONAGRACEAE (EVENING-PRIMROSE FAMILY)
Fig. 5-59
Flowering (February–)March–April(–July)

Field recognition: Stemless winter annual or biennial herb from a taprooted basal rosette of elliptic to reverse lance-shaped, irregularly feather-lobed leaves. Flowers showy, with four yellow petals, opening near sunset and arising at ground level in the center of the rosettes. Seed capsules reverse pyramid-shaped, very hard, 1–2 cm long with four wings near the top, maturing near ground at the base of plant. Of moist lowlands, open flats, and grassy areas; often a weed of lawns.

Similar species: None. (But compare its stemless growth habit with that of the other evening-primroses in the Ottine Wetlands.)

Remarks: These yellow wildflowers grow in moist, grassy, disturbed areas, including lawns, where they are considered weeds by some and are notable for their nearly stemless condition, with the flowers arising directly from the center of the rosette of leaves. When they open near sunset, nectar-seeking hawk-moths (Sphingidae) such as the lined sphinx (*Hyles lineata*) visit them.

Distribution: They occur from Kentucky, Tennessee, and northern Alabama west to Kansas and south through the Blackland Prairie and westward in Texas.

Fig. 5-59
Stemless evening-primrose (*Oenothera triloba*).

Pigeon berry
Rivina humilis C. Linnaeus
PHYTOLACCACEAE (POKEWEED FAMILY)
Fig. 5-60
Flowering (March–)June–October

Field recognition: Usually low, perennial herb with spreading branches from a
thick rootstock and bright green, stalked, egg-shaped to elliptic-egg-shaped
or diamond-shaped, long-pointed leaves that are broadly rounded to straight
at the base, and with margins often undulate. Flowers greenish, white, rose to
reddish-purple in upper axillary or terminal open spikes. Fruits red or orange
berries, each containing a hairy seed. Of shaded, moist alluvial woods and
thickets as well as shaded limestone slopes in scrub or chaparral.

Similar species: None.

Remarks: Like those of the related pokeweed (*Phytolacca americana*) (Fig.
3-19 [p. 79]), its leaves, roots, and fruits are reported to be poisonous (Diggs,
Lipscomb, and O'Kennon 1999), though apparently not to birds that likely
consume the brightly colored berries and disperse the seeds.

Distribution: Throughout the southeastern United States through Central,
South, and West Texas to tropical America.

Fig. 5-60
Pigeon berry (*Rivina
humilis*).

Red-seed plantain
Plantago rhodosperma J. Decaisne
PLANTAGINACEAE (PLANTAIN FAMILY)
Fig. 5-61
Flowering (March–)April–May

Field recognition: Taprooted annual
herb, stemless with a rosette
of basal, reverse lance-shaped,
parallel-veined, smooth-margined
to distantly toothed or lobed, hairy
gray-green leaves. Flowers with
membranous, translucent whitish
or yellowish petals, bracted (each
above a reduced leaf), and borne in
erect, long-stalked spikes. Bright
red to reddish-black seeds developing in capped, 2-seeded capsules enclosed by the translucent
petals and sepals. Seeds sticky-
mucilaginous when wet. Of various
habitats, upland sandy to rocky
soils and alluvial to disturbed
sandy or gravelly soils of washes,
streambeds, and pond margins.

Fig. 5-61
Red-seed plantain (*Plantago rhodosperma*).

Similar species: Pale-seed plantain (*P. virginica*) (flowering March–May[–June])
differs in having yellow-brown to black rather than reddish seeds. It is a
species of similar rocky, sandy, or gravelly soils in the eastern third of Texas
and the eastern United States. It has spread westward as a weed with human
disturbance (e.g., roadsides, fields) (March–May[–June]).

Remarks: Red-seed plantain has very small white, chaffy flowers. It may grow
in gravelly spots along creek beds, though its usual habitation is more upland
than lowland.

Distribution: From the lower Mississippi River valley south and west through
the western two-thirds of Texas to Arizona; mostly the Blackland Prairie and
westward.

Drummond's phlox
Phlox drummondii W. Hooker
POLEMONIACEAE (PHLOX FAMILY)
Fig. 5-62
Flowering spring

Field recognition: Pubescent annual herb to 0.5 m tall with narrow, opposite
sessile leaves. Flowers in small, dense inflorescences, with a short tube (2 cm)
and five intense red petal limbs and a darker red "eye" in the center. Capsular

fruit containing small seeds. Of neutral to acidic sandy soils in open woodlands and grasslands.

Similar species: None.

Remarks: The rich red flower of Drummond's phlox has made it, according to some historical accounts, "the most famous Texan in the world." It can be seen in spring along roadsides and on higher ground near the marshes and swamps of the wetland itself.

Distribution: Drummond's phlox was once an endemic of East and Southeast Texas, occurring naturally nowhere else in the world. Now its many cultivars are distributed across the globe.

Fig. 5-62
Drummond's phlox (*Phlox drummondii*).

Swamp smartweed
Polygonum hydropiperoides A. Michaux
POLYGONACEAE (KNOTWEED FAMILY)
Fig. 5-63
Flowering June–November

Field recognition: Annual or perennial, smooth herb to over 1 m tall with erect to sprawling stems often rooting at the swollen nodes (points of leaf attachment). Leaves lance-shaped with the base of each forming a tubular, membranous sheath (ochrea) enclosing the stem for a short distance above the node and terminating in bristlelike cilia to 5 mm long. Flowers produced in dense terminal or nearly terminal, sometimes branching spikelike clusters 4–8 cm long. Tepals (petals/sepals) white to pink. Fruit dark brown or black, shiny, 3-sided seeds. Of low, wet areas throughout Texas.

Fig. 5-63
Swamp smartweed (*Polygonum hydropiperoides*).

Similar species: Hairy or bristle smartweed (*P. setaceum* W. Baldwin) (Fig. 5-64), to over 2 m in height, is a similar species of similar habitats in the south-

eastern and southwestern United States, East and Southeast Texas, and the Edwards Plateau that reaches its
southwestern limit in the state in the Ottine Wetlands
(Turner et al. 2003a). It is distinguished by its wider leaf
blades (10–60 mm versus 8–15 mm wide) and ochrea
with cilia 6–22 mm long. Pennsylvania smartweed
(*P. pensylvanicum* C. Linnaeus), a widespread species
throughout North America and Texas, is an annual to
1.5 m in height, with often cherry-red stem nodes, erect
pink inflorescences, and ochrea that usually lack cilia.

Remarks: Swamp smartweed's small white to pink flowers
appear in marshes and other wet habitats, where the
plants grow thickly enough to form mats (May–June).

Distribution: Occurring widely throughout North America and south into the Tropics and beyond.

Fig. 5-64
Bristle smartweed
(*Polygonum setaceum*).

Heart sorrel
Rumex hastatulus W. Baldwin
POLYGONACEAE (BUCKWHEAT FAMILY)
Fig. 5-65
Flowering April–June

Field recognition: Slender annual
or short-lived perennial herb to
two-thirds meter in height, with
lance-shaped to narrowly oblong,
mostly basal, pale green leaves
that have two projecting, earlike
teeth near the base of the blades
(hastate). Inflorescences slender,
narrowly branched, and unisexual,
with different plants producing staminate and pistillate flowers (dioecious). Flowers greenish-yellow
with orange or reddish stigmas or
anthers, and dangling in whorls
from short, threadlike stalks. Small
and somewhat heart-shaped with
six tepals (petals and sepals), the
inner three loosely adhering to
and enclosing the 3-sided seeds in
pistillate plants. Of open, seasonally moist, sandy areas.

Similar species: None.

Fig. 5-65
Heart sorrel (*Rumex hastatulus*).

Remarks: The red and yellow, heart-shaped flowers bloom in open, sandy, disturbed ground surrounding the wetlands proper. Some people eat the leaves despite the presence of potentially poisonous oxalic acid, albeit in small amounts. The taste has been described as a "pickle-sour" flavor.

Distribution: Also known as "Engelmann's dock," this plant ranges throughout the southeastern Coastal Plain north to North Carolina and inland to Illinois, Missouri, and Oklahoma and through the eastern half of Texas. May occur as a waif in ruderal areas northward.

Coastal brookweed
Samolus ebracteatus K. Kunth subsp.
alyssoides (A. A. Heller) R. Knuth
PRIMULACEAE (PRIMROSE FAMILY)
Fig. 5-66
Flowering (March–)April–October

Field recognition: Low, hairless, light green perennial (or annual?) herb with semi-succulent, entire reverse egg-shaped to spoon-shaped or reverse lance-shaped, alternate leaves rounded at tips, and tiny white, 5-petaled, bell-shaped flowers in long- and stout-stalked, open spikes. Fruits many-seeded globose capsules. Of sandy soils along streams, creeks, ditches, and seepage areas, often in brackish or saline marshes and flats near the coast.

Similar species: Thin-leaf brookweed (*S. parviflorus* C. Rafinesque-Schmaltz) is very similar and occurs in the same habitats in the Ottine Wetlands but has shorter, sessile or almost stalkless, lateral spikes of slightly smaller flowers with a shorter petal tube ([March–]April–September[–October]). It ranges across North America from southern Canada to tropical America.

Remarks: Coastal brookweed is a white-flowered wetland species that grows in seepage areas and along stream banks. It is notable for its tolerance of saline soil.

Fig. 5-66
Coastal brookweed
(*Samolus ebracteatus*).

Distribution: Subspecies *alyssoides* occurs along the coastal southeast from Florida through the West Indies to Texas and Mexico and inland to the Ottine Wetlands.

Ten-petal thimbleweed
Anemone berlandieri G. Pritzel
RANUNCULACEAE (BUTTERCUP FAMILY)
Fig. 5-67
Flowering late February–mid-April

Field recognition: Clumped perennial herb from a tuberous root. Leaves basal, 3-parted, and composed of sessile, roughly egg-shaped scalloped-toothed to shallowly cleft, rounded lobes. Flowers borne on a long stalk above a leaf-like involucre of finely divided, whorled bracts (reduced leaves) well above the basal leaves. Flowers of 10–20 petal-like sepals, white within and blue-lavender to violet outside, surrounding numerous pollen-bearing stamens and pistils (ovaries). Fruiting head an elongate, cylindrical spike of numerous, single-seeded achenes covered with long white, woolly hairs. Of calcareous or sandy clay soils in open grasslands and prairies and as a weed in lawns.

Similar species: Carolina anemone or windflower (*A. caroliniana* T. Walter), a species of drier, usually sandy soils of the eastern Great Plains and sandy barrens and woodlands of the southeastern United States through East and Southeast Texas to the Edwards Plateau, differs in being rhizomatous or stoloniferous (spreading by underground or ground-level stems) and having finely divided basal leaves and an involucre below rather than above the middle of the flowering stalk.

Remarks: Ten-petal thimbleweed bears white or blue to violet flowers.

Distribution: This plant prefers clay soils whether sandy or limestone enriched, and occurs in interior prairies of the southeastern United States and south-central Great Plains through East, Central, and Southeast Texas.

Fig. 5-67
Ten-petal thimbleweed
(*Anemone berlandieri*).

Blue larkspur
Delphinium carolinianum T. Walter
RANUNCULACEAE (BUTTERCUP FAMILY)
Fig. 5-68
Flowering March–June(–July)

Field recognition: Slender perennial herb from short, fleshy roots with simple stems to 1 m high. Leaves deeply fan-dissected into fine, branching, linear segments. Flowers spurred and borne in simple spikes on short, erect stalks, appearing, especially in bud, like tiny, breaching dolphins returning to the water. Dark, rich blue to white with five petal-like sepals, the upper forming a long spur that contains the spurred upper two petals that secrete nectar into their horn. Fruit three fused pods that split along their upper inner margins to release many tiny (1–2 mm) seeds via shaking. Of sandy or clayey soils in prairies, woodland openings, and pine woods.

Similar species: None.

Remarks: Blue larkspur need not be blue at all, as illustrated by the white flower shown in the photograph. Despite its appearance in the disturbed habitat surrounding the oxbow lake of Palmetto State Park, it is generally characterized as a plant of dry uplands and prairies.

Distribution: Blue larkspur ranges throughout the central Great Plains and interior southeastern United States south to northern and central Mexico.

Fig. 5-68
Blue larkspur
(*Delphinium
carolinianum*); white
variety from oxbow
lake region.

Showy buttercup
Ranunculus macranthus G. Scheele
RANUNCULACEAE (BUTTERCUP FAMILY)
Fig. 5-69
Flowering late March–early May(–June)

Field recognition: Perennial herb with divided, mostly basal leaves composed of 3–5 deeply lobed leaflets having long, erect, spreading hairs on the stems and leafstalks. Flowers with 10–22 large (12–22 mm long), bright yellow, showy petals. Fruits laterally flattened, beaked, single-seeded achenes aggregated into dense, headlike clusters in the flower centers. Of moist woods, flood-plains, swamps, seepage slopes, and riverbanks.

Similar species: Tufted buttercup (*R. fascicularis* G. H. Muhlenberg *ex* J. Bigelow) (Fig. 5-70) is a widespread, earlier-flowering (late February–early April) species of the eastern United States through East and Southeast Texas that is adapted to slightly drier, sandy habitats such as sandy open woods and moist meadows and prairies. It may be distinguished by its fewer (five, rarely to nine), smaller (7–12 mm long) petals, and less hairy stems and leafstalks with appressed or inclined, rather than erect, spreading hairs.

Remarks: The yellow flowers of showy buttercup may be seen in periodically wet terrain of Palmetto State Park. It is probably an important food for the American oil beetle (*Meloë americanus*), which was a notable find during our studies here.

Distribution: As currently interpreted (Whittemore 1997), showy buttercup is an endemic of east-central Texas. Populations formerly included in this species occurring in West Texas and southeastern Arizona are now placed in *R. fasciculatus* M. de Sesse y Lacasta & J. M. Mociño (not to be confused with *R. fascicularis*), a member of the *R. petiolaris* K. Kunth *ex* A. P. de Candolle group native to northern Mexico.

Fig. 5-69
Showy buttercup (*Ranunculus macranthus*).

Fig. 5-70
Tufted buttercup (*Ranunculus fascicularis*).

Weak buttercup

Ranunculus pusillus J. Poiret
RANUNCULACEAE (BUTTERCUP FAMILY)
Fig. 5-71
Flowering March–April(–June)

Field recognition: Pale green annual, wetland herb with smooth stems and leafstalks and egg-shaped to oblong, long-stalked basal leaves. Stem leaves simple,
  entire to slightly toothed, reverse lance-shaped, narrowly elliptic, or lance-
  shaped to linear. Flowers small with few (1–3) tiny (1.5–2.0 mm long) petals.
  Fruits small, smooth, single-seeded achenes with beak weakly developed or
  absent. Of shallow water, mud in seepage areas, ditches, ponds, marshes, and
  swamps.
Similar species: The reduced petals and annual growth habit of this buttercup
  are distinctive.
Remarks: This is one of three buttercup species seen in the Ottine Wetlands. Its
  small yellow flowers are "often inconspicuous and easily overlooked" (Diggs,
  Lipscomb, and O'Kennon 1999).
Distribution: From the northeastern United States south and west through
  Missouri and the southeast to East and Southeast Texas and the Edwards
  Plateau of Central Texas; also occurs in coastal northern California.

Fig. 5-71
Weak buttercup
(*Ranunculus pusillus*).

White avens
Geum canadense N. von Jacquin
ROSACEAE (ROSE FAMILY)
Fig. 5-72
Flowering April–May(–June)

Field recognition: Slender, rhizomatous perennial herb with long-stalked, basal, feather-compound leaves composed of three large terminal, diamond-shaped, saw-toothed to lobed leaflets and 2–4 much smaller, lower ones that are sometimes lacking. Stem leaves sharply toothed, nearly stalkless to sessile, 3-parted to undivided. Flowers solitary on long stalks at the ends of a few spreading branches of the upper stem. Five petals, white fading to yellowish, surrounding numerous stamens and central ovaries. Fruit a spherical head of single-seeded achenes with long (4–7 mm) style beaks. Of rich woods and lowland thickets.

Similar species: None. (But compare Canada sanicle [*Sanicula canadensis*], especially in fruit.)

Remarks: White avens is a white-petaled herb that prefers swamps, bogs, and "rich woods" (Correll and Johnston 1970; Correll and Correll 1972).

Distribution: Through the eastern half of the United States and the eastern half of Texas, excluding the southern part of the state, and is at the southern margin of its range in the Ottine Wetlands.

Fig. 5-72
White avens
(*Geum canadense*).

Prairie gerardia
Agalinus heterophylla (T. Nuttall) J. K. Small *ex* N. Britton
SCROPHULARIACEAE (SNAPDRAGON FAMILY)
Fig. 5-73
Flowering June–October

Field recognition: Bushy, annual, hemiparasitic herb with dark green stems and thick, rigid, narrow leaves ranging from 3-cleft and broadly linear lower down through narrowly linear to awl-like higher up on the plant. Flowers showy, foxglovelike, deep pink to lavender-tinged white with two yellow lines in the purple-brown spotted throat. Fruit a many-seeded globose capsule. Of moist grasslands, prairies, or open woodland.

Similar species: Purple gerardia (*A. purpurea* (C. Linnaeus) F. Pennell) differs in having linear to awl-like leaves that are never cleft, and in having very short sepal lobes much shorter than the petal tube, rather than as long as or longer than the tube as in prairie gerardia. Of moist, sandy soils in the southeastern United States to East and Southeast Texas. Blooming from August to November.

Remarks: In Texas this purple and white flower occurs in the south-central part of the state but not in the north-central region (Diggs, Lipscomb, and O'Kennon 1999). Perhaps this is explained by its association with wetlands. It is a species of bogs and seepage areas though not restricted to such wet habitats.

Distribution: Throughout the eastern United States through East and Southeast Texas.

Fig. 5-73
Prairie gerardia
(*Agalinus heterophylla*).

Fig. 5-74
Coastal water-hyssop
(*Bacopa monnieri*).

Coastal water-hyssop
Bacopa monnieri (C. Linnaeus) F. Pennell
SCROPHULARIACEAE (SNAPDRAGON FAMILY)
Fig. 5-74
Flowering April–September

Field recognition: Low, mat-forming perennial, succulent herb, pale green with fleshy, smooth-margined, spoon-shaped leaves and bell-shaped white to pale blue or lilac flowers borne on short stalks in the axils (angles) of the leaves. Fruits small, ovoid, many-seeded capsules. Of mud or sand in or along margins of ponds, streams, and ditches.
Similar species: None.
Remarks: This is a true wetland species with typically bluish flowers and fleshy leaves and stems.
Distribution: Coastal water-hyssop grows in mats in the sand and mud of ponds and streams throughout the southeastern United States west through East to Central and South Texas.

Texas paintbrush
Castilleja indivisa G. Engelmann
SCROPHULARIACEAE (SNAPDRAGON FAMILY)
Fig. 5-75
Flowering (March–)April–May(–June)

Field recognition: Hairy annual herb from a short taproot, to one-third meter tall. Leaves linear-lance-shaped with one or two short, diverging lobes. Flowers short, greenish-yellow, and tubular, enclosed by bright red-orange-tipped leaves (bracts) and sepals in a short, congested, round-topped, terminal spike. Capsular fruits containing many small seeds.

Similar species: None.

Remarks: Texas paintbrush is one of the most familiar and readily recognized wildflowers of its namesake state. Its red leaves and lighter-colored flowers may be seen growing in disturbed areas alongside the oxbow lake of Palmetto State Park and in the pastures that must be crossed to reach certain swamps and marshes, but it is not a species of the wetland proper. This plant is believed to be parasitic on the roots of other plants and is possibly poisonous as well (Diggs, Lipscomb, and O'Kennon 1999).

Distribution: Texas paintbrush has a limited distribution, occurring only in southeastern Oklahoma and the eastern half of Texas.

Fig. 5-75
Texas paintbrush (*Castilleja indivisa*).

Clasping false pimpernel
Lindernia dubia (C. Linnaeus) F. Pennell
SCROPHULARIACEAE (SNAPDRAGON FAMILY)
Fig. 5-76
Flowering late May–October

Field recognition: Small, open-branching annual herb growing cryptically amid other vegetation in mud, sand, or shallow water along stream, pond, and wetland margins. Leaves small, opposite, egg-shaped to elliptic, 3–5-veined and nearly sessile. Flowers arising singly from leaf axils on long pedicels, small, tubular, snapdragon-like with a 2-lipped mouth, pale blue, lavender, or white. Fruit small capsules.

Similar species: Narrow-leaf conobea (*Leucospora multifida* (A. Michaux) T. Nuttall) (Fig. 5-77) is a related (snapdragon family) low annual species of similar wetland to seasonally moist habitats and weedy in disturbed areas with tiny pink to pale lavender, tubular 2-lipped flowers (June–October). It is easily distinguished by its low bushy habit and once or twice feather-divided, finely glandular-hairy foliage. We observed it growing at the trampled, muddy margin of a cattle pond along the north fork of Rutledge Creek. Narrow-leaf conobea occurs throughout the eastern half of the United States and eastern half of Texas.

Remarks: The flowers of clasping false pimpernel vary in color but typically have a lavender hue. Unlike its relative the Texas toad-flax, clasping false pimpernel is a true wetland species of swamp mud and stream bank.

Fig. 5-76
Clasping false pimpernel (*Lindernia dubia*).

Fig. 5-77
Narrow-leaf conobea (*Leucospora multifida*).

Distribution: Throughout most of the United States south to South America and East and Southeast Texas west to the Great Plains and Edwards Plateau.

Texas toad-flax

Nuttallanthus (Linaria) texanus
(G. Scheele) D. Sutton
SCROPHULARIACEAE (SNAPDRAGON FAMILY)
Fig. 5-78
Flowering February–May

Field recognition: Annual or winter biennial herb with erect flowering stems to two-thirds meter high and sparse, narrow, linear leaves. Flowers borne in erect, terminal spikes, pale lavender-blue, each appearing flat and upturned with a reflexed upper and a projecting lower lip, over 1 cm long with a thin, down-curved, backward-pointing spur 5–9 mm in length. Fruit an ovoid capsule, splitting from the top to release many minutely, densely warty seeds via shaking. Of seasonally moist, sandy soils of fields, open woods, and pinelands, mostly in the eastern two-thirds of Texas.

Fig. 5-78
Texas toad-flax (*Nuttallanthus texanus*).

Similar species: Old field toad-flax (*N. (L.) canadensis* (C. Linnaeus) D. Sutton) has smaller flowers that are less than a centimeter in length with a 2–6-mm long spur and smooth or sparsely tuberculate seeds. It occurs rarely in the eastern third of the state but is widespread throughout the United States and southern Canada. In fact, Texas toad-flax has been considered a variety of that species (as *L. canadensis* (C. Linnaeus) C. Dumortier var. *texana* (G. Scheele) F. Pennell).

Remarks: Texas toad-flax has light violet petals and a preference for open, sandy woods rather than the lower wetlands of swamp and marsh.

Distribution: From coast to coast in the United States and southern Canada, though sporadic in the Southwest.

Downy ground-cherry
Physalis pubescens C. Linnaeus
SOLANACEAE (POTATO FAMILY)
Fig. 5-79
Flowering April–November

Field recognition: Pale green, sticky-hairy annual herb of low, moist, often disturbed ground with egg-shaped, smooth-margined to weakly and irregularly toothed leaves. Flowers hanging from short, curved stalks from leaf axils (angles), bell-shaped and pale yellow with a brownish-purple throat or "eye." Pollen-bearing anthers blue or violet. Fruit a small berry enclosed in an inflated, top-shaped, green husk.

Similar species: Clammy ground-cherry (*P. heterophylla* C. Nees von Esenbeck) is very similar in appearance but is a perennial from a deep rootstock and has yellow anthers, though sometimes blue or violet-tinged. This species blooms from April to October. Of sandy, often disturbed areas in the eastern United States and adjacent Canada and the eastern half of Texas.

Fig. 5-79
Downy ground-cherry
(*Physalis pubescens*).

Beach ground-cherry (*P. cinerascens* (M. Dunal) A. Hitchcock) (flowering April–October) is another perennial that differs from the previous two in having dense, stellate, nonglandular (not sticky) hairs on its leaves. Of drier sandy soils and disturbed areas throughout Texas, northern Mexico, and the south-central United States. Field ground-cherry (*P. mollis* T. Nuttall), of doubtful distinction from the latter species, has felty, soft-pubescent leaves. It is said to bloom from May to July.

Remarks: Downy ground-cherry is also known as "husk-tomato."

Distribution: Across the southern United States and the eastern half of Texas south through Mexico, the West Indies, and Central America.

Buffalo bur
Solanum rostratum M. Dunal
SOLANACEAE (POTATO FAMILY)
Fig. 5-80
Flowering (April–)June–October

Fig. 5-80
Buffalo bur (*Solanum rostratum*).

Field recognition: Prickly annual, weakly woody, low shrub with once- or twice-feather-lobed leaves. Flowers yellow with drooping, flasklike stamens opening by pores at their long tips, surrounding a downward-curving style that points either left or right (enantiostylous). Fruit a yellowish berry enveloped in a dry, prickly, husklike calyx. An aggressive weed of overgrazed pastures, fields, and waste areas throughout the state.

Similar species: Use of the common name "mala mujer" instead of buffalo bur invites confusion with the euphorb *Cnidoscolus texanus*, which prefers similar habitats and also has prickly stems, but white rather than yellow flowers.

Remarks: Buffalo bur is a poisonous plant with yellow flowers, a preference for the higher, drier soil between the wetlands and the pastures that overlook them, and a prickly fruit that is dispersed on the fur of mammals. The "handedness" or enantiostylous condition of the flowers, occurring as they do in two mirror-image forms, is believed to be a mechanism promoting cross-pollination among different plants.

Distribution: Buffalo bur's natural range lies on the Great Plains from Texas to Nebraska, but its weedy nature has sent it farther east and west.

Texas nightshade
Solanum triquetrum A. Canavilles
SOLANACEAE (POTATO FAMILY)
Fig. 5-81
Flowering throughout the year

Field recognition: Perennial sprawling to arching and twining shrub from a woody base and roots. Stems green and angled. Leaves triangular-heart-shaped to 3-lobed. Flowers starlike, in few-flowered, dangling clusters with white to blue-tinged petals and yellow anthers that are arranged in the form of a cone. Fruits small red berries. Of stream banks, thickets, slopes, floodplains, and disturbed areas.

Similar species: None.

Remarks: We saw the dangling white, star-shaped flowers of Texas nightshade along trails in Palmetto State Park where the uncommon iridescent emerald-green Haldeman's potato beetle (*Leptinotarsa haldemani*) was mating and feeding on its foliage. The anthers shed pollen through pores at their tips and are adapted for pollination by pollen-collecting bees. The bees grasp them, and while dangling upside down, vibrate their wing muscles, causing pollen to be released onto their undersides, from which they comb it and pack it into pollen "baskets" on their hind legs.

Distribution: Throughout Central, South, and West Texas and adjacent Mexico.

Fig. 5-81
Texas nightshade (*Solanum triquetrum*).

Bog-hemp
Boehmeria cylindrica (C. Linnaeus) O. Swartz
URTICACEAE (NETTLE FAMILY)
Fig. 5-82
Flowering June–October

Field recognition: Tallish perennial herb (to 1.5 m) with pale green, lance-shaped, elliptical to broadly egg-shaped leaves having rounded-saw-toothed to finely scalloped margins. Flowers small, nonshowy, unisexual in axillary (in the angles of the leaves) spikes, the pollen-bearing, staminate spikes of interrupted clusters, pistillate spikes continuous, sometimes on different plants. Fruit a tiny, flattened, ovoid, slightly beaked, seed-containing achene surrounded by a corky wing.

Fig. 5-82
Bog-hemp (*Boehmeria cylindrica*).

Similar species: The related stinging or heart-leaf nettle (*Urtica chamaedryoides* F. Pursh) is an annual herb to less than 1 m in height with stinging hairs and flowers in short, dense, globose clusters in the leaf axils. It ranges through the southeastern United States west through East and Southeast Texas to the West Cross Timbers and Edwards Plateau. Pennsylvania pellitory (*Parietaria pensylvanica* G. H. Muhlenberg *ex* C. von Willdenow) (Fig. 5-83) is another smaller (to two-thirds meter) stingless relative (family Urticaceae) of similar or drier habitats that has thin, hairy, entire egg-shaped to narrow-elliptic or lance-shaped leaves and sessile (stalkless), axillary flower clusters surrounded by green involucral bracts that are divided into narrow lobes. It occurs throughout temperate North America. Florida pellitory (*P. floridana* T. Nuttall) has rounded to triangular leaves and tastes of cucumber (Williams and Watson 1978). It occurs in widely scattered localities throughout the southeastern United States, the West Indies, Mexico, and South America, including an apparently isolated locality in the Ottine Wetlands.

Remarks: Bog-hemp's habitat requirements have been characterized in prose perfect for a wetland species, for it occurs in "bogs, marshes, swamps, seepage areas and in wet soil and water along rivers and

Fig. 5-83
Pennsylvania pellitory (*Parietaria pensylvanica*).

streams" (Correll and Johnston 1970). We observed a male giant walkingstick (*Megaphasma dentricus*) as it consumed a bog-hemp leaf near Rutledge
Swamp. This is the first record of the plant as a suitable food for the insect.

Distribution: Throughout the eastern half of the United States west into the
Great Plains and scattered westward; in Texas, the eastern and southeastern
portions of the state to the western Edwards Plateau and Rolling Plains.

Lance-leaf frog-fruit
Phyla (Lippia) lanceolata (A. Michaux) E. Greene
VERBENACEAE (VERVAIN FAMILY)
Fig. 5-84
Flowering May–October

Field recognition: Perennial herb of moist or wet areas from creeping, prostrate
woody stems (stolons) rooting at the nodes with ascending, erect shoots with
square stems. Leaves opposite, bright green, lance-shaped, elliptic-lance-
shaped to egg-shaped with conspicuous feather-venation, saw-toothed above
the middle and tapered below to a wedge-shaped base. Inflorescences in globose to cylindrical heads (elongating in fruit) on elongate stalks. Composed
of purplish-brown scales surrounding very small pale blue, purplish, or white
flowers with yellow throats ("eyes") that turn orange-red with age.

Similar species: Common frog-fruit (*P. (L.) nodiflora* (C. Linnaeus) E. Greene)
(Fig. 5-85) of low, moist, often disturbed areas, and a cosmopolitan weed of
subtropical and tropical regions of the world, is a plant of smaller stature with
smaller, narrower, semi-succulent or fleshy leaves spoon- or wedge-shaped to
reverse lance-shaped, and toothed near the tips. Wedgeleaf frog-fruit (*P. (L.)
cuneifolia* (J. Torrey) E. Greene), reported recently for this area by Williams
and Watson (1978), is a species of low grasslands, plains, and prairies of the

Fig. 5-84
Lance-leaf frog-fruit (*Lippia lanceolata*).

Fig. 5-85
Common frog-fruit (*Lippia (Phyla) nodiflora*).

western half of Texas, the southwestern United States, and northern Mexico that reaches its eastern limit here. It has shorter, narrower leaves, larger scales subtending the flower heads (4–5 mm versus 2–3 mm) on shorter stalks that only just exceed the adjacent leaves.

Remarks: Lance-leaf frog-fruit is a bona fide wetland plant as its common name implies, with blue, purplish, or white flowers.

Distribution: Along river bottoms, stream banks, shores, ponds, and marshes from the eastern half of the United States throughout Texas (except the Trans-Pecos portion of the state) to northern Mexico; also California.

Vervains

Harsh, Sandpaper vervain
Verbena scabra M. H. Vahl
Gulf vervain *V. xutha* J. Lehmann
VERBENACEAE (VERVAIN FAMILY)
Figs. 5-86, 5-87
Sandpaper vervain: Flowering March–December
Gulf vervain: Flowering (March–)June–October

Field recognition: Sandpaper vervain: Large, rough, open-branching perennial herb of marshes, swamps, pond margins, streamsides, and other low, wet areas, with stiff, stalked, egg-shaped, toothed leaves very rough (sandpapery) above. Small blue, pink, or lavender, 5-lobed, tubular flowers borne in slender, elongate, terminal spikes. Fruits enclosed by the sepals and consisting of four single-seeded nutlets.

Fig. 5-86
Harsh, sandpaper vervain (*Verbena scabra*).

Field recognition: Gulf vervain: Coarse, tufted, pubescent perennial herb from a rootstock with square stems and sessile oblong to egg-shaped, coarsely toothed, 3-cleft to feather-cut leaves. Blue to purple flowers borne in stiff, elongate spikes. Of stream bottoms and low to dry, sandy disturbed areas. Across the southern United States from Alabama to Arizona and throughout Texas.

Similar species: White or nettle-leaf vervain (*V. urticifolia* C. Linnaeus), reported by Bogusch (1928) and Parks (1935) and supported by collection records (Turner et al. 2003a), is a species of bottomland woods and streamsides in the eastern United States and west through East and Southeast Texas

Fig. 5-87
Gulf vervain (*Verbena xutha*).

to north-central Texas and the Edwards Plateau. It differs from sandpaper vervain in having longer-pointed, more coarsely scalloped-saw-toothed leaves similar to those of a nettle (*Urtica*), that are neither stiff nor roughly sandpapery above, and in having white flowers (June–October).

Remarks: These two species have bluish flowers, though the color is often not strong.

Distribution: One standard reference noted that the authors had seen few harsh vervain specimens from Texas localities (Diggs, Lipscomb, and O'Kennon 1999), despite its occurrence from coast to coast within the United States (Correll and Johnston 1970). Gulf vervain is more limited in its U.S. range, occurring from Alabama to Arizona. Neither species is a wetland plant per se.

Missouri violet
Viola missouriensis E. Greene
VIOLACEAE (VIOLET FAMILY)
Fig. 5-88
Flowering mid-March–mid-April(–May)

Field recognition: Stemless, short-rhizomatous perennial herb with large basal, hairless, rounded, egg-shaped to long-triangular, long-pointed leaves with heart-shaped to straight bases. Flowers solitary, nodding on long, erect, basal stalks, pale blue with white centers bordered with dark violet and with dark violet veins on the lower, spurred, and lateral petals. Five petals, the lower spurred, containing nectar. Fruits 3-celled, valved, green capsules with purple spots containing many orange-brown to yellowish-brown seeds bearing oily appendages (elaiosomes). Of shaded stream bottom woods and floodplain forests.

Similar species: Sister violet (*V. sororia* C. von Willdenow), of which Missouri violet is sometimes considered a variety (*V. s.* var. *missouriensis* (E. Greene) L. McKinney), may be distinguished by its broader, more egg-shaped-triangular leaves (about as wide as long versus longer than wide) with shorter-tipped leaves that are long-hairy on their undersurfaces. Introgressive hybridization does occur, however, between these two entities, and intermediate individuals may be expected where they co-occur.

In sandy soils of open woods and weedy in disturbed areas, even cities, throughout the eastern half of the United States and southern Canada through the eastern half of Texas.

Primrose-leaf violet (*V. primulifolia* C. Linnaeus), a species of the northeastern United States and Canada south and west through Indiana, Oklahoma, East Texas, and reaching its southwestern, disjunct limit here in the Ottine Wetlands, was reported by both Bogusch (1928) and Parks (1935a), and their reports are supported by collection records (Turner et al. 2003a). Primrose-leaf violet is quite distinctive in having egg-shaped to

Fig. 5-88
Missouri violet (*Viola missouriensis*).

lance-egg-shaped, long-stalked leaves with a tapering, wedge-shaped base and white flowers with light blue veins, especially on the lower, spurred petal.

Remarks: We found Missouri violet near Central Texas creeks, including Copperas Creek of the upland Lost Pines forest and Rutledge Creek of the lowlying Ottine Wetlands.

Distribution: Missouri violet occurs from the eastern Great Plains south to Southeast Texas and west to the Trans-Pecos and New Mexico.

MONOCOTS

The monocotyledonous or lilylike wildflowers are those flowering plants that are not quite grasslike but do have narrow, elongate leaves with parallel veins and flower parts in multiples of three or six. Several rare species of orchids, lilies, irises, and yellow-eyed-grasses have disjunct, outlying, or marginal populations in the Ottine Wetlands, which represent the westernmost localities for their species. These include the water spider orchid (*Habenaria repens*), Carolina spider lily (*Hymenocallis caroliniana*), purple fleur-de-lis (*Iris hexagona* var. *flexicaulis*), rose pogonia (*Pogonia ophioglossoides*), scythe-fruit arrowhead (*Sagittaria lancifolia*), Carolina yellow-eyed-grass (*Xyris caroliniana*), and iris-leaf yellow-eyed-grass (*X. laxifolia* var. *iridifolia*). In addition, the south-western helleborine orchid (*Epipactis gigantea*); the South Texas and Mexican spotted American-aloe (*Manfreda maculosa*); and the coastal, South Texas, and northeastern Mexican green lily (*Schoenocaulon drummondii*) each reach their respective limits or are disjunct here. A notable ornamental introduction of some concern is the yellow flag (*Iris pseudacorus*), as it competes for space with the native purple fleur-de-lis.

Scythe-fruit arrowhead
Sagittaria lancifolia C. Linnaeus
ALISMATACEAE (ARROWHEAD FAMILY)
Fig. 5-89
Flowering May–November

Field recognition: Tall, robust, aquatic or paludal (marsh), coarse-rhizomatous perennial herb to nearly 2 m in height. Leaves emergent, long, with lance-shaped, elliptic to broadly egg-shaped blades to 40 cm long by 10 cm wide. Flower stalks branched-whorled, often branching at the lower nodes. Fruits developing on ascending stalks, heads of flattened, beaked, single-seeded achenes. Of coastal, brackish tidal marshes, freshwater marshes, and swamps and along streams inland.

Similar species: Burheads (*Echinodorus* spp.) differ from the arrowheads (*Sagittaria* spp.) in having only bisexual flowers and

Fig. 5-89
Scythe-fruit arrowhead (*Sagittaria lancifolia*).

fruiting heads of plump, ribbed, wingless, erect-beaked achenes that give the head a "burred" (short-spiny) appearance. Erect burhead (*E. berteroi* (K. Sprengel) N. Fassett) has erect branched-whorled inflorescences that branch from the lower nodes, whereas creeping burhead (*E. cordifolius* (C. Linnaeus) A. Grisebach) has arching to sprawling inflorescences that root and produce plantlets from the nodes. The former is widespread throughout Texas, the central United States, and Mexico; the West Indies to South America; also California. The latter occurs throughout the southeastern United States, the eastern half of Texas, the West Indies, and Mexico to South America. Both grow in habitats similar to those of arrowheads.

Remarks: Scythe-fruit arrowhead is a tall wetland plant of marshes and swamps and, according to one source, it might not be expected so far from the Gulf

Coast (Diggs, Lipscomb, and O'Kennon 1999). It has white flowers and a leaf that is more spindle-shaped than arrow-shaped, despite its common name.

Distribution: From the coastal Atlantic and Gulf United States through the West Indies, Mexico, and Central America to South America; disjunct inland to the Ottine Wetlands.

Delta arrowhead
Sagittaria platyphylla (G. Engelmann) J. G. Smith
ALISMATACEAE (ARROWHEAD FAMILY)
Fig. 5-90
Flowering April–October

Field recognition: Aquatic or marshland, stoloniferous (spreading by aboveground stems, or stolons), perennial herb from corms that have milky sap. Leaves with parallel main veins, submerged, floating, or emergent. Submerged and floating leaves lax, straplike, and without blades. Emergent leaves with egg-shaped, elliptic or lance-shaped blades to 8 cm wide. Flowers unisexual, in whorls of three on emergent stalks, staminate above and pistillate below, each with three white petals. Fruits developing on down-curved stalks from the lower, pistillate flowers. Small,

Fig. 5-90
Delta arrowhead (*Sagittaria platyphylla*).

flattened-globose heads of tiny flattened, beaked, single-seeded achenes. Of mud and shallow water in marshes, swamps, ponds, and sloughs.

Similar species: Grass-leaf arrowhead (*S. graminea* A. Michaux) (Figs. 5-91, 1-19 [p. 21]) is a coarse-rhizomatous, non-stoloniferous (spreading only via underground stems, or rhizomes) perennial sister species that differs in having the emergent leaves bladeless or with linear or narrowly lance-shaped to linear-reverse lance-shaped blades to 2.5 cm wide and fruits borne on ascending rather than down-curved stalks. Of similar habitats throughout the eastern half of the United States through East and south-central Texas. It is at the southwestern margin of its range here.

Remarks: Delta arrowhead is an aquatic plant with whorls of white, 3-petaled flowers. It is a fixture of swamps and marshes and may be seen in profusion, along with its grass-leaved relative, at the "extinct mud-boil" pond on the Hiking Trail in Palmetto State Park.

Fig. 5-91
Grass-leaf arrowhead
(*Sagittaria graminea*).

Distribution: Delta arrowhead occurs throughout the deep central-southern United States (scattered northeastward) through East and Southeast Texas to the Rolling Plains and Edwards Plateau.

Green dragon
Arisaema dracontium (C. Linnaeus)
H. Schott
ARACEAE (ARUM FAMILY)
Fig. 5-92
Flowering late April–May(–June)

Field recognition: Perennial herb from corm (a swollen underground stem) with a single large, compound leaf of 11–15 elliptic to reverse lance-shaped, feather-veined leaflets of varying size, with the veins curving and joining along the leaflet margins. The leaf is basally thrice-divided, with the two outer portions further subdivided into 5–7 leaflets of descending size. These are arranged on the outside of the two curved, lateral axes, the whole forming a ringlike arrangement. Inflorescence a thick spike (spadix) with a tapering, pale yellow to orange terminal, nonflower-

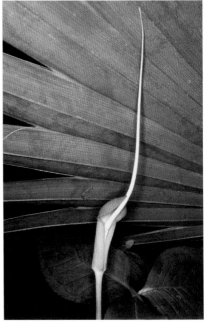

Fig. 5-92
Green dragon (*Arisaema dracontium*) with backdrop of dwarf palmetto (*Sabal minor*) frond.

ing portion up to 15 cm long known as an "appendix" that functions as an insect attractant. Unisexual flowers borne near the base, the pollen-bearing, staminate above the pistillate in middle-aged plants and enclosed by a light green, inrolled leaf (spathe) with a slender, pointed tip. Fruits reddish-orange berries. Of rich woodlands, thickets, and alluvial or floodplain forests, often colonial.

Similar species: None.

Remarks: Green dragon is a close relative of Jack-in-the-pulpit (*Arisaema triphyllum* (C. Linnaeus) H. Schott) with green inflorescences and red fruits. Small or young plants have only staminate flowers and are functionally male. Plants of intermediate size and age are monoecious (having flowers of both sexes), whereas the oldest and largest individuals possess only pistillate flowers and are thus functionally female (Clay 1993). Green dragon is poisonous and if eaten will cause burning and swelling of the mouth and throat (Diggs, Lipscomb, and O'Kennon 1999). The cells contain needlelike crystals of calcium oxalate called "raphides" that lacerate tissues and cause intense burning and swelling. This is a defense against herbivores.

Distribution: Through the eastern United States and southern Canada west to Wisconsin and through East and Southeast Texas to the East Cross Timbers and Edwards Plateau.

Erect dayflower
Commelina erecta C. Linnaeus
COMMELINACEAE (SPIDERWORT FAMILY)
Fig. 5-93
Flowering May–June(–July–August–)September–October

Field recognition: Perennial, clumped herb from tuberous (fleshy) roots, with erect to sprawling stems and linear to lance-egg-shaped leaves sheathing the stem nodes at their bases and with earlike lobes at the summits of the sheaths. Flowers borne in dense clusters enclosed by a folded, boat-shaped, leafy spathe that is fused along the back or basal margin. Flowers short-lived, emerging one by one daily, or a few days apart, from the spathe. Usually

Fig. 5-93
Erect dayflower
(*Commelina erecta*).

Fig. 5-94
Common dayflower
(*Commelina communis*).

wilting and liquefying by noon. The two upper, stalked petals blue, standard or bannerlike, and showy. Lower petal tiny and white. Three upper stamens sterile, modified into nectar-containing knobs. Of diverse habitats, moist to dry, and soils from clayey and silty to sandy and rocky. Occasionally a weed of disturbed areas or cultivated ground.

Similar species: Spreading dayflower (*C. diffusa* N. L. Burnman) was reported by both Bogusch (1928) and Parks (1935a) from the Ottine Wetlands. It is an annual or perennial herb of moist, clayey soils and low floodplain woods as well as weedy in disturbed waste and cultivated areas of the southeastern United States through East and Southeast Texas (flowering April–November). It is easily distinguished by its spathe, which is open along the entire margin, and by the flower, the lower petal of which is also blue and only somewhat smaller than the upper two. Common or Asiatic dayflower (*C. communis* C. Linnaeus) (Fig. 5-94), an adventive from cultivation and native to Asia, occurs in low, moist areas and stream banks and is a weed of cultivated and disturbed areas of the eastern half of the United States. It has an open spathe like that of *C. diffusa* but a reduced, pale lower flower petal like that of *C. erecta*.

Remarks: This blue-petaled flower appears along stream banks and in thickets in the wild. It also sprouts as a weed in gardens.

Distribution: Erect dayflower ranges throughout the southern two-thirds of the United States east of the Rocky Mountains and throughout Texas south to Central America.

False dayflower
Tinantia anomala (J. Torrey) C. B. Clarke
COMMELINACEAE (SPIDERWORT FAMILY)
Fig. 5-95
Flowering April–June(–July)

Field recognition: Annual herb with erect, clustered stems and broadly egg-shaped to lance-shaped pale green, powdery leaves sheathing the stem nodes at their bases, heart-shaped and clasping the stem at the summit of the sheath. Inflorescence an elongate cluster surrounded below by a flat (not folded), egg-shaped, leaflike spathe. Flowers with two large, showy, non-stalked, diamond-shaped, lavender-blue upper petals, with the lower, third one much smaller and white. Six stamens, densely bearded with purple and yellow-tipped hairs, the upper lateral two looking somewhat like a pair of eyes in the center of a floral "face." Of gravelly, rocky limestone soils, usually shaded and also weedy in moist, disturbed areas, wood edges, and ravines, more or less mesic, at least seasonally (spring).

Similar species: The single flat spathe below the inflorescences separates false dayflower from *Tradescantia* sp. (with two spathes) and *Commelina* sp. (with a folded one).

Remarks: Also known as "widow's tears," this blue flower is not limited to wetlands. It has a preference for coarse soils and shade.

Distribution: False dayflower is nearly endemic to the Edwards Plateau of Central and west-central Texas with a single record from Durango, Mexico (Diggs, Lipscomb, and O'Kennon 1999).

Fig. 5-95
False dayflower
(*Tinantia anomala*).

Giant spiderwort
Tradescantia gigantea J. Rosa
COMMELINACEAE (SPIDERWORT FAMILY)
Fig. 5-96
Flowering (February–)March–May

Field recognition: Perennial, somewhat succulent herb with mucilaginous sap, to 1 m tall, yellow-green with narrow, elongate, linear-lance-shaped leaves sheathing the swollen stem nodes at their bases. Stem internodes and upper leaves with minute, dense, velvety, nonglandular pubescence. The flowers are borne in congested, terminal clusters surrounded by a pair of leafy bracts that are distinctly swollen or sac-shaped at their bases. Flowers with three magenta-pink to blue petals and three sepals covered with fine, dense, velvety hairs. Flowers emerging singly early in the morning of each day, wilting and liquefying by afternoon. Of limestone and clayey soils of pasturelands, thickets, and floodplain terrace woods in Palmetto State Park. The plants are gregarious and colonial.

Similar species: Ohio spiderwort (*T. ohiensis* C. Rafinesque-Schmaltz) occurs in sandy to clayey soils of meadows, thickets, and roadsides throughout the eastern United States and is common and widespread throughout the eastern half of Texas. It may be distinguished by its smooth, powdery blue-green leaves, and stems and sepals that lack the fine, velvety pubescence of those of giant spiderwort.

Remarks: The purple flowers of spiderworts are sometimes mistaken for those of the purple iris, as we discovered in conversation with a visitor to Palmetto State Park in early spring. Ohio spiderwort is the most common spiderwort in the eastern United States, and it hybridizes with many other species where they co-occur. In fact, according to Faden (2000) most *Tradescantia* spp. of Texas hybridize freely where they co-occur. The filaments of the stamens are

Fig. 5-96
Giant spiderwort
(*Tradescantia
gigantea*).

clothed in long purple hairs in which Robert Brown first observed, under a microscope in 1828, cytoplasmic (cell sap) streaming.

Distribution: Giant spiderwort is an endemic species of the eastern Edwards Plateau, the southwestern part of East Texas, and scattered localities in north-central Texas and is possibly introduced in Louisiana (Foster 2000). Astride the swamps and marshes of Gonzales County it reaches its southernmost limit (Turner et al. 2003b).

Herbertia
Herbertia lahue (J. Molina) P. Goldblatt
IRIDACEAE (IRIS FAMILY)
Fig. 5-97
Flowering mid-March–early May

Fig. 5-97
Herbertia (*Herbertia lahue*).

Field recognition: Iridoid (*Iris*-like), long-stalked perennial herb from bulb. Often colonial in clayey to sandy soils of open woodlands, grasslands, prairies, and lawns. Basal leaves long, linear, pleated, and sheathing at the bases. Stem leaves shorter and entirely sheathing the nodes. Stems terminating in an inflorescence composed of 1–5 flowers surrounded by a sheathing leaf or spathe. Flowers essentially *Iris*-like but smaller, the three outer tepals (petals/sepals) largest, spreading, dark or light lavender spotted with a violet pattern at base surrounded by a violet halo. Inner three tepals much smaller and dark violet to blackish. Style 3-forked, the branches with forked, pollen-receptive stigmas at their tips. Pollen-bearing anthers narrow, slender, and lying against the undersides of the style branches. Fruit a 3-parted capsule with a flat (truncated) top, developing below the tepals (inferior).

Similar species: Celestial-lily or prairie-pleatleaf (*Nemastylis geminiflora* T. Nuttall), a related bulbose iridoid of open clayey and limestone-derived soils of open woodlands and grasslands, was reported for these wetlands by Parks (1935a). Of the southern midwestern, central-south, and southeastern Great Plains (Tallgrass Prairie) through East and Southeast Texas west to the Rolling Plains and Edwards Plateau. Vegetatively similar, it may easily be distinguished by its six sky-blue tepals, all similar in size and shape.

Remarks: We found this attractive blue-purple flower growing in large numbers on a lawn across the parking lot from the refectory in Palmetto State Park during mid-April. The flower opens in the morning, and if a colony is revisited after dark by flashlight, no flowers will be seen, for they close, wilt, and begin to liquefy by midafternoon. The plant, like the fleur-de-lis, is a member of the iris family.

Distribution: It is endemic through Southeast and East Texas and coastal Louisiana and is introduced in Florida; also South America (Argentina, Brazil, Chile, and Uruguay), where four other species occur.

Yellow flag
Iris pseudacorus C. Linnaeus
IRIDACEAE (IRIS FAMILY)
Fig. 1-14 [p. 19]
Flowering April–May

Field recognition: Perennial rhizomatous herb with leaves in two rows, folded lengthwise and overlapping at the base in a fanlike arrangement (equitant), nearly a meter or more in length. Forming thick, extensive colonies in swamps and along the shores of rivers and lakes. Flowering stems to 1.5 m high, bearing inflorescences surrounded by pairs of green, leaflike spathes. Flowers large, bright yellow, and showy. Petals of three deflexed or spreading "falls," three erect to spreading "standards," with the style divided into three petal-like branches, each overlying a stamen and the base of a fall. Ovary below petals; fruit a 3-celled capsule containing many seeds.

Similar species: Purple fleur-de-lis or Dixie iris (*I. hexagona* T. Walter var. *flexicaulis* (J. K. Small) R. Foster) (Fig. 1-15 [p. 19]) is a native of extreme southeastern Texas and Louisiana and is disjunct in the Ottine Wetlands (Correll and Johnston 1970; Turner et al. 2003b). It is easily recognized by its smaller stature and blue-purple to violet flowers (flowering March–May) with spreading standards. It has recently been considered a hybrid of two other species: short-stem iris (*I. brevicaulis* C. Rafinesque-Schmaltz) (flowering April–June) of the Midwest, deep south-central, and extreme East Texas; and giant blue iris (*I. giganticaerulea* J. K. Small) (flowering March–April) of Alabama, Louisiana, and Mississippi (Henderson 2002). If this interpretation is accurate, then its correct scientific name is *I.* X *flexicaulis* J. K. Small (*I. brevicaulis* X *I. giganticaerulea*). Carolina iris (*I. hexagona* T. Walter *sensu strictu* (var. *hexagona*)) occurs only in the extreme southeastern United States (South Carolina, Georgia, and Florida).

This purple fleur-de-lis is the native iris and the unofficial floral symbol of the Ottine Wetlands. Much more abundant is the alien yellow iris (*I. pseudacorus*), which seems to be pushing the native aside in competition for the same (or similar) ecological niche. We looked for the purple species during two flowering seasons, but we found it only once, in late April, and only two plants were in bloom, though the exotic was flowering by the dozens nearby. The native may well appear earlier than the adventive (March–May versus

April–May), but in our experience it appeared more than one week *later* than the yellow iris.

Remarks: Yellow flag is an exotic iris native to the Old World continents of Europe and Africa. It has escaped cultivation in the New World and is now flourishing in the Ottine swamps alongside a correspondingly reduced population of native purple fleur-de-lis.

Distribution: Thirty years before our study began, the standard guide to the vascular plants of Texas reported that yellow flag was established in a single county, Hardin County (325 km northeast of Ottine), and nowhere else in the state (Correll and Johnston 1970). Their characterization of the attractive species as "rather aggressive" is borne out by its dominating presence in lagoons along the Palmetto Trail. A more recent distribution map reports this species from five counties widely separated from one another in the eastern half of the state (Turner et al. 2003b). It is now scattered throughout North America (Flora of North America Editorial Committee 2002b).

Dotted blue-eyed-grass
Sisyrinchium pruinosum E. Bicknell (*S. langloisii* E. Greene)
IRIDACEAE (IRIS FAMILY)
Fig. 5-98
Flowering March–May

Field recognition: Low, perennial, clumped herb from fleshy, fibrous roots with equitant (sheathing and in two rows, like *Iris*), powdery blue-green, short, narrow, sword-shaped leaves. Numerous stems bearing small clusters (umbels) of flowers enclosed by a pair of leaflike bracts (spathe) from which they emerge one by one daily, and from which the globose seed capsules dangle. Flowers of six, minutely pointed tepals, violet to purplish-blue with a yellow "eye," rarely white.

Fig. 5-98
Dotted blue-eyed-grass (*Sisyrinchium pruinosum*).

Similar species: Sword-leaf blue-eyed-grass (*S. ensigerum* E. Bicknell (*S. chilense* W. Hooker)) was recently reported for these wetlands by Williams and Watson (1978). A prairie species of the western edge of the Blackland Prairie to South and West Texas and southern Oklahoma, it is known to hybridize with *S. pruinosum* where they co-occur, as they do here. They may be distinguished by the minutely pubescent ovaries of *S. ensigerum* versus smooth in *S. pruinosum;* however, intermediates may be expected (Correll and Johnston 1970; Cholewa and Henderson 2002). Narrow-leaf or Bermuda blue-eyed-grass (*S. angustifolium* P. Miller) was also reported by Parks (1935a) for the area; this appears to be supported by more recent collection records (Turner 2003b). A species of low, moist habitats throughout the eastern United States, it reaches its southern limit in the Ottine Wetlands. Narrow-leaf blue-eyed-grass has a rather distinctive flowering habit, with a very long outer floral bract, much longer than the inner, and the stems bearing the inflorescences are flattened, with "wings" on each side over 1 mm wide, or greater than the stem width.

Remarks: Dotted blue-eyed-grass is not a grass but an iris relative, and it is not limited to wetlands. Its blue, purple, or violet flower also appears as a weed on lawns.

Distribution: Dotted blue-eyed-grass grows in clay and sandy-clay soils of prairies, open woods, disturbed areas, roadsides, and lawns through the south coastal United States and Deep South (Mississippi River valley) west through East and Southeast Texas to the West Cross Timbers and Edwards Plateau.

Wild onion
Allium canadense C. Linnaeus
LILIACEAE (LILY FAMILY)
Fig. 5-99
Flowering March–May

Field recognition: Perennial herb from a subterranean bulb with an onion or garlic odor and long, slender, narrowly linear basal leaves with tubular sheathing bases. Flowers numerous, fragrant, borne in long-stalked, round-topped, umbrella-shaped clusters, often replaced by or intermixed with sprouting bulblets. Six tepals (undifferentiated petals/sepals), white, pink, or lavender when present. Fruits are capsules, splitting open at maturity to reveal shiny black seeds. Of various habitats, usually moist, sandy, or rocky soils of open woods, meadows, prairies, and roadsides.

Similar species: Drummond's or prairie onion (*A. drummondii* E. von Regel) (Fig. 5-100) is a middle and southern Great Plains species ranging throughout Texas, mostly in drier limestone-derived soils from Nebraska south throughout Texas to northern Mexico (flowering March–May[–June]). It may be distinguished by noting that the perianths (tepals) are urn-shaped and permanently enclose the seed capsule at maturity, rather than wilting away from it as in wild onion.

Fig. 5-99
Wild onion (*Allium canadense*).

Fig. 5-100
Drummond's onion (*Allium drummondii*).

Pink-flowered Drummond's onion honors a Scottish botanist who collected plants in the United States after the Revolutionary War. Many native Texas plants bear his name.

Remarks: Wild onion has white or pale pink flowers and grows in a variety of habitats. Its common name suggests edibility, but when eaten in quantity by cattle, the plant can be lethal. Children have become sick as well (Diggs, Lipscomb, and O'Kennon 1999). However, we tried it (in moderation) with no ill effects.

Distribution: Also known as "Canada garlic," the species occurs in the eastern half of the United States and through East, Southeast, and north-central Texas to the Edwards Plateau.

Rain-lily
Cooperia drummondii W. Herbert (*Zephyranthes chlorosolen* (W. Herbert) D. Dietrich)
LILIACEAE (LILY FAMILY)
Fig. 5-101
Flowering (May–)June–October

Field recognition: Perennial, often colonial herb from subterranean bulbs with narrow, linear leaves (1–3 [rarely 5] mm wide) in spring. Single, long-stalked flower following rains of summer through fall. Flower with six white tepals (petals and sepals) fused into a long, narrow tube (8–18 cm long). Fruit a 3-celled globose capsule containing neatly stacked black, flattened, wedge-shaped, triangular seeds. Flowers expanding in late afternoon or evening a day or two following heavy rains. May last up to four days before wilting. Of usually thin, dry soils over limestone.

Similar species: Giant rain-lily (*C. pedunculata* W. Herbert (*Z. drummondii* D. Don)) (Fig. 5-102) flowers earlier in the year ([March–]April–July[–August]). The larger, short-pedicelled flowers with a short tube (2.2–4.0 cm long) on thicker stalks appear around dusk a few days following heavy rains, expand overnight, and, turning pale pinkish, wilt within one or two days. The strap-shaped, powdery blue-green leaves (4–12 mm wide) are also distinctive. Of moister, usually calcareous soils of prairies, roadsides, and open woods of Southeast, Central, and South Texas and adjacent Mexico.

Remarks: Rain-lilies flower profusely in late afternoons and evenings following the downpours of summer and fall. From this alone one might correctly predict that the white-flowered plants are not confined to wetlands, nor do they prefer sandy soils over those rich in calcium.

Fig. 5-101
Rain-lily (*Cooperia drummondii*).

Fig. 5-102
Giant rain-lily (*Cooperia pedunculata*).

Distribution: Throughout the south-central United States from Kansas through north-central, Southeast, Central, and South Texas south into Mexico.

Copper-lily
Habranthus tubispathus (L'Heritier de Brutelle) H. Traub
LILIACEAE (LILY FAMILY)
Fig. 5-103
Flowering July–October following rains

Field recognition: Perennial herb from small bulbs with single, long-stalked flowers following rains of summer and fall. Flowers funnel-shaped, orange-yellow with reddish tinge distally and on the outer surface of the tepals (petals/sepals), inclined and somewhat 2-lipped. Fruit a 3-lobed, nearly globose capsule. Leaves basal, narrow linear to strap-shaped, following flowers. Of moist, open areas and a weed of lawns.

Similar species: None.

Remarks: The yellow-orange flower of copper-lily appears, like that of the white-flowered rain-lily, after rains in summer and fall.

Fig. 5-103
Copper-lily (*Habranthus tubispathus*).

Distribution: One authority has the species listed as a Texas endemic of the eastern half of the state (Correll and Johnston 1970), whereas a later treatment reports at least a few western counties within the plant's range (Turner et al. 2003b), and another relays a report that copper-lily might have been introduced from South America (Diggs, Lipscomb, and O'Kennon 1999); if so, it is not a Texas endemic at all, but an introduced exotic.

Crow-poison
Nothoscordum bivalve (C. Linnaeus) N. Britton
LILIACEAE (LILY FAMILY)
Fig. 5-104
Flowering March–early May, again late September–October(–December)

Field recognition: Onionlike but odorless perennial herb from a subterranean bulb with long, slender, narrowly linear basal leaves, tubular at their bases and sheathing the bulb neck. Flowers nonfragrant and borne in long-stalked, few-flowered clusters, with six white to cream-colored tepals (petals/sepals) with yellow bases inside, and purple-red or greenish midribs outside. Six stamens. Fruits 3-lobed, membranous capsules, splitting open at maturity to reveal black, angled, or flattened seeds. Of open, low, and sandy woods, prairies, and disturbed areas throughout Texas.

Similar species: Also known as "false onion" or "false garlic," crow-poison is sometimes placed in the same genus (as *Allium bivalve* C. Linnaeus). However, it lacks the characteristic sulfurous onion odor (allyl sulfides) of the other *Allium* spp.

Remarks: Crow-poison grows in a variety of soil types, appears as a weed in urban lawns, and is one of the most abundant plants native to Texas. It is not apparently poisonous.

Fig. 5-104
Crow-poison (*Nothoscordum bivalve*).

Distribution: Through the southeastern quarter of the United States, throughout Texas to New Mexico, Arizona, Mexico, and south to South America.

Water-hyacinth
Eichhornia crassipes
PONTEDERIACEAE (PICKERELWEED FAMILY)
Fig. 5-105
Flowering (April–)late June–September

Field recognition: Usually free-floating, perennial, aquatic herb of lakes, ponds, slow-moving rivers, canals, and ditches. Leaves leathery, shiny-green, egg-shaped to kidney-shaped in floating rosettes with swollen, spongy, inflated stalks. Spreading vegetatively via slender, elongate lateral stems. Inflorescences elongating overnight with flowers opening within two hours after sunrise the following day or two and wilting by night. By the following day the flower stalk has bent over, submerging the developing fruits underwater, where the seeds mature. Flowers very showy, blue or lavender-blue with a dark blue blotch surrounding a yellow "eye" on the upper petal. Tristylous (styles of three different lengths in different flowers, bearing the pollen-receptive stigmas at different heights, a mechanism promoting outcrossing) in their native Amazonia (Horn 2002). Roots fibrous, sometimes rooted in mud along water margin.

Similar species: None.

Remarks: This Brazilian native is a floating aquatic plant with splendid purple flowers. However, because it clogs waterways and competes with natives, it is typically described as a "beautiful, noxious weed" (Correll and Johnston 1970). In fact, it is illegal to *possess* water-hyacinth in Texas (Diggs, Lipscomb, and O'Kennon 1999). On the beneficial side, it does remove pollutants from water and has been utilized in sewage treatment ponds for this very purpose.

Distribution: From the Atlantic Ocean to Texas and south into its tropical homeland. Native to Brazilian Amazonia, where populations exhibit tristyly, or three flower morphs that vary in style length and anther position. This system promotes cross-fertilization among the different flower types. Introduced along the southern and southeastern coastal United States, California, and throughout the Tropics and subtropics worldwide, where it aggressively spreads vegetatively. Needless to say, this includes coastal South and East Texas and scattered localities inland.

Fig. 5-105
Water-hyacinth (*Eichhornia crassipes*).

6
Ferns and Horsetails

Despite a small sample size of a dozen or so species, the ferns and related horsetails nevertheless display the same distribution patterns shown by the other floral components examined thus far. Narrow-leaf and Virginia chain ferns (*Woodwardia areolata* and *W. virginica,* respectively) both make this their approximate southwesternmost outpost, as does the cinnamon fern (*Osmunda cinnamomea*). The southwestern limits of southern shield fern (*Thelypteris kunthii*) and bracken (*Pteridium aquilinum* var. *pseudocaudatum*) east of Trans-Pecos Texas occur nearby as well. Engelmann's adder's tongue fern (*Ophioglossum engelmannii*) and blunt-lobed wood fern (*Woodsia obtusa*) are near their southern limits here. A disjunct, southwestern collection record for the eastern fragile fern (*Cystopteris fragilis* var. *protrusa*) has also been published (Turner et al. 2003b).

Mosquito fern
Azolla caroliniana C. von Willdenow
AZOLLACEAE (MOSQUITO FERN FAMILY)
Fig. 6-1
Sporulating summer–fall

Field recognition: Tiny, free-floating perennial fern (surviving through the year in temperate areas) forming small, velvety, deep green to reddish mats on the surface of quiet waters or along muddy or mucky margins. Stems to 1 cm long with deeply 2-lobed, overlapping (imbricate) leaves and, except for the roots hanging below, resembling a small leafy liverwort. Spores of two types (heterospory); megaspores producing female gametophytes and microspores that grow into sperm-producing male gametophytes. On still waters of ponds, lakes, and sluggish streams and stranded on mud.

Similar species: No free-floating *leafy* liverwort occurs in our area; moreover, its roots identify mosquito fern as a vascular plant rather than a bryophyte. Its leafy stems and the fine-velvety texture of its upper, floating leaves will serve to distinguish it from duckweeds (*Spirodela* and *Lemna* spp.), tiny flowering plants that are stemless and have thalli with smooth, shiny upper surfaces.

Fig. 6-1
Mosquito fern
(*Azolla caroliniana*);
sporophytes.

Remarks: Mosquito fern is a small aquatic species that might be confused with a liverwort. Its hollow, upper leaf lobes are home to a symbiotic cyanobacterium (*Anabaena azollae* Strasburger) that converts ("fixes") atmospheric nitrogen (N_2 gas) into forms (nitrates) that the plant can use as food. As a result the little plants are sometimes used, like legumes, for fertilizer ("green manure") and cattle feed. Mosquito fern may turn red under stress, and large mats covering pond surfaces provide a striking spectacle (Diggs, Lipscomb, and O'Kennon 1999).

Distribution: Throughout the eastern half of the United States and Texas through Mexico and the West Indies to South America (Patagonia); also Europe and Asia.

Narrow-leaf chain fern
Woodwardia areolata (C. Linnaeus) T. Moore
BLECHNACEAE (CHAIN FERN FAMILY)
Fig. 6-2
Sporulating March–November

Field recognition: Perennial herbaceous fern from slender, long-creeping rhizomes with fronds of two types. Fertile fronds with narrow, reduced, bladeless lobes and discrete, linear-oblong sori in chainlike rows along each side of their midveins. Indusia flaplike, opening next to midvein, permitting the sporangia to protrude. Sterile fronds feather-lobed, egg-shaped-oblong, composed of reverse lance-shaped lobes with slightly wavy and finely saw-toothed margins. Of wet, usually sandy or acidic, peaty areas such as bogs, low woods, swamps, marshes, thickets, bases of seepage slopes, and along streams.

Similar species: None.

Remarks: Narrow-leaf chain fern is now considered a close relative of Virginia chain fern (*W. virginica*) (Fig. 6-3 [p. 209]), though at one time it was placed in a different genus (as *Lorinseria areolata* (C. Linnaeus) K. Presl). Its fertile

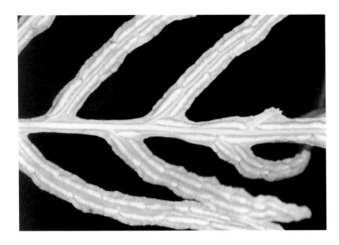

Fig. 6-2
Narrow-leaf chain fern (*Woodwardia areolata*); fertile sporophyte frond.

fronds bear spore-producing bodies in chainlike rows on each side of the midribs of the fertile fronds.

Distribution: This species grows in marshes, swamps, and seeps and occurs from extreme southeastern Canada and the southeastern United States south and west through Arkansas and Missouri to East and Southeast Texas to a southwestern limit in the Ottine Wetlands.

Virginia chain fern
Woodwardia virginica (C. Linnaeus) J. E. Smith
BLECHNACEAE (CHAIN FERN FAMILY)
Fig. 6-3
Sporulating April–December

Field recognition: Large, coarse perennial (evergreen?) fern from stout, woody, creeping-elongate, black rhizomes with large (to over 1.3 m long), leathery, dark green fronds of one type (sterile and fertile fronds similar in appearance), feather-compound, oblong-elliptic to elliptic-lance-shaped, composed of elongate feather-lobed leaflets with smooth margins. Veins of leaflet lobes anastomosing (joining after dividing) to form rows of oblong areoles (rings or loops) on each side of the midveins. Fruit dots (sori) oblong-linear and arranged in chainlike rows on either side of the midveins of the leaflet lobes in place of the areoles. Flaplike indusia opening on the midvein side, permitting protrusion of mature sporangia and release of spores. In moist or wet, sandy or peaty soils of bogs, seepage slopes, stream banks, swamps, and low, shrubby thickets.

Similar species: None.

Remarks: This fern is recognized by the remarkable end-to-end pattern of spore-producing bodies (sori: sporangia clusters or "fruit dots") on the lower surface of its fertile fronds, which may be compared to the similar pattern of

Fig. 6-3
Virginia chain fern
(*Woodwardia
virginica*); fertile
sporophyte frond.

its close relative the narrow-leaf chain fern (Fig. 6-2 [p. 208]). We encountered the Virginia species along seepage areas and stream banks and in swamps.

Distribution: From the northeastern United States and southeastern Canada south and west along the Atlantic and Gulf Coastal Plain through East Texas and disjunct to southeast-central Texas to its southwestern limit in the Ottine Wetlands.

Bracken

Pteridium aquilinum (C. Linnaeus) F. Kuhn
var. *pseudocaudatum* (W. Clute) A. A. Heller
DENNSTAEDTIACEAE (BRACKEN FAMILY)
Fig. 6-4
Sporulating June–November

Field recognition: Coarse colonial, perennial, deciduous fern from long-creeping, subterranean stems with ascending, spreading-horizontal, thrice-divided fronds to over 1 m high, broad-triangular, each division feather-divided into elongate, feather-lobed leaflets with elongate terminal leaflets. Fruit dots (sori) marginal and covered by the inrolled margins of the ultimate leaflet lobes (a false indusium). Usually in sandy soils of dryish, upland areas such as open woods, edges of woods, and thickets and pastures, often colonizing burnt-over areas.

Similar species: None.

Remarks: Bracken is perhaps the best known of all ferns because it occurs throughout the world in suitable habitats, such as the sandy, forested regions of east-central Texas. Its spore-bearing bodies (sporangia clustered into sori) are arranged along the margins of the lower surface of the fronds. Consumption of this fern can lead to serious illness and has been associated with stomach cancer (Diggs, Lipscomb, and O'Kennon 1999).

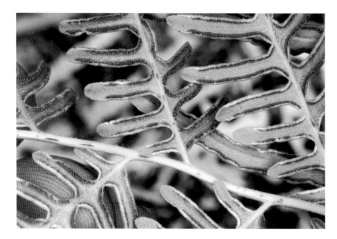

Fig. 6-4
Bracken (*Pteridium aquilinum*); fertile frond showing marginal sporangia.

Distribution: The "tailed" var. *pseudocaudatum,* with distinctive elongate terminal leaflets, ranges throughout the southeastern United States west through East and Southeast Texas to its southwestern limit in nearby Wilson County. Not until the mountains of Trans-Pecos Texas does the species reappear—as another variety (*pubescens* L. Underwood). The species, including all its varieties, is nearly worldwide in distribution from tropical to boreal latitudes.

Cinnamon fern
Osmunda cinnamomea C. Linnaeus
OSMUNDACEAE (CINNAMON FERN FAMILY)
Fig. 6-5
Sporulating March–July (and later)

Field recognition: Large, coarse perennial fern to over 1.5 m tall from a short, woody, creeping rhizome. Fronds of two types. Inner circle of fertile fronds developing first, narrow with reduced, bladeless leaflets densely covered with cinnamon-colored sporangia and soon wilting. Sterile, photosynthetic fronds developing later in an outer ring, feather-compound with feather-lobed leaflets, oblong-lance-shaped to elliptic-lance-shaped, to a meter or more in length. In wet, acidic soils of swamps, marshes, seepage slopes, lake edges, and bogs.
Similar species: None.
Remarks: This is the giant fern of the Ottine Wetlands reaching heights of 1.5 m or more. We found it only in the wet, mucky terrain of Rutledge Swamp.
Distribution: Cinnamon fern, a nonflowering plant named for the color of its spore-bearing structures, occurs from the eastern United States and southern Canada to East and Southeast Texas as far west as the Ottine Wetlands; also New Mexico (Correll and Johnston 1970) and an isolated record from Uvalde County in west-central Texas (Turner et al. 2003b).

Fig. 6-5
Cinnamon fern
(*Osmunda
cinnamomea*).

Southern shield fern
Thelypteris kunthii (N. Desvaux) C. Morton
THELYPTERIDACEAE (MARSH FERN FAMILY)
Figs. 6-6–6-8
Sporulating April–November

Field recognition: Herbaceous perennial fern from a creeping rhizome with large (to 1.3 m long), pubescent, light green, feather-compound, egg-shaped-oblong to oblong or elliptic-lance-shaped fronds of long, narrow, feather-lobed leaflets. Fruit dots (sori) round, located along lateral veins of the leaflet lobes between the midvein and margins, and covered by round to kidney-shaped leafy flaps (indusia). Of sandy stream and creek banks, swamps, low woods, and seepages at the base of limestone bluffs.

Similar species: None.

Remarks: We show three photographs of southern shield fern to illustrate the life history stages of ferns in general: the familiar sporophyte (spore-producing) stage growing in South Soefje Swamp (Fig. 6-6); the spore-producing

Fig. 6-6
Southern shield fern (*Thelypteris kunthii*); sporophyte growing in South Soefje Swamp.

Fig. 6-7
Southern shield fern; sporophyte frond showing
clustered sporangia ("fruit dots" or sori).

Fig. 6-8
Southern shield fern; gametophytes.

bodies (sporangia clustered into sori) on the lower surface of a sporophyte
frond (Fig. 6-7); and the small, liverwort-like, heart-shaped prothallus, the
seldom-noticed gametophyte stage that grows from germinating spores and
produces the next generation of sporophytes by sexual reproduction (Fig. 6-8).

Distribution: Southern shield fern occurs from the southeastern United States
and Caribbean through East and Southeast Texas; scattered westward to
north-central Texas and the Edwards Plateau and southwest to nearby Wilson County.

Smooth horsetail
Equisetum laevigatum A. Braun
EQUISETACEAE (HORSETAIL FAMILY)
Fig. 6-9
Sporulating May–July

Field recognition: Rhizomatous evergreen perennial with rough, longitudinally
finely grooved, hollow, jointed green stems. Leaves inconspicuous, reduced to
rings of scales just above stem nodes ("joints"). Sporangia produced in small
terminal cones. Spores of one size (homosporous). Of moist or wet, sandy
soils along streams, lakes, seepage slopes, alluvial thickets, marshes, wet
meadows, and prairies.

Similar species: Tall scouring-rush (*E. hyemale* C. Linnaeus) has thicker, more
robust stems with conspicuous dark rings above the nodes. Sporulating from
late March to late fall, it ranges throughout Texas and North America and
also Eurasia.

Remarks: Smooth horsetail is a nonflowering, cone-bearing, primitive plant that reaches a height of nearly 1.5 m. It has a life cycle similar to that of ferns. Its leaves are small, scalelike, and most easily recognized by their tooth-like tips. Another common name is "scouring rush" because of the silica-impregnated stems that were once used to scrub pots and pans. The abrasive surface may be felt by running the edge of a thumbnail over the stem surface perpendicular to its long axis. These plants are reputed to be poisonous, like water-hemlock. We saw a single specimen, the individual pictured here from Rutledge Creek, and were in fact expecting the closely related species *E. hyemale* instead, a widespread species throughout North America and Eurasia reported from

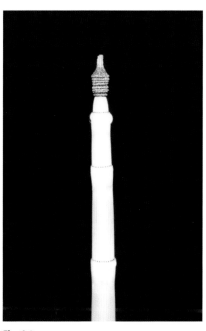

Fig. 6-9
Smooth horsetail (*Equisetum laevigatum*).

Gonzales County (Turner et al. 2003b). The latter typically bears dark bands on the stem that were not visible on the plant we stumbled upon. It bears cones from late March through late fall.

Distribution: This species occurs from southern Canada throughout most of the United States except the Southeast. In Texas it occurs in the central, west, north-central and northern portions of the state. It also occurs in Mexico and Guatemala (Correll and Johnston 1970).

7
Duckweeds, Bryophytes, and Algae

Duckweeds and watermeals (family Lemnaceae) are highly reduced, aquatic flowering plants (monocots) that rarely produce flowers. For purely practical purposes they are treated here with bryophytes and algae, which are giants by comparison. They have no recognizable leaves or stems and, rarely, produce submicroscopic flowers reduced to individual stamens and pistils.

Bryophytes include the mosses, liverworts, and hornworts. Thirty-one species of mosses, fourteen liverworts, and one hornwort have been recorded from the Ottine/Palmetto State Park area. Many members of this ancient and highly vagile (via spore dispersal) group have scattered, worldwide distribution patterns, occurring on several different continents. Such is the case with most of those that occur here, though their ranges within North America are usually eastern or southeastern with their western limits in Central Texas. These include the mosses *Archidium hallii* Aust., *Astomum ludovicianum* (Sull.) Sull., *A. muhlenbergianum* (Sw.) Grout, *A. phascoides* (Hook. *ex* Drummond) Grout, *Atrichum angustatum* (Brid.) B. S. G., *Barbula cancellata* C. Muell., *Clasmatodon parvulus* (Hampe) Hook. & Wils. *ex* Sull., *Desmatodon plinthobius* Sull. & Lesq. *ex* Sull., *Ditrichum pallidum* (Hedw.) Hampe, *Entodon seductrix* (Hedw.) C. Muell., *Steerecleus* (*Rhynchostegium*) *serrulatum* (Hedw.), *Fissidens bushii* (Card. & Ther.) Card. & Ther., *F. bryoides* Hedw., *Hygroamblystegium tenax* (Hedw.) Jenn., *Leskea australis* Sharp, *Leucobryum albidum* (Brid. *ex* P. Beauv.) Lindb., *Leucodon julaceus* (Hedw.) Sull., *Philonotis longiseta* (Michx.) Britt., *Pogonatum brachyphyllum* (Michx.) P. Beauv., *Sematophyllum adnatum* (Michx.) Britt.; and the liverworts *Asterella tenella* (L.) P. Beauv., *Cephalozia lunulifolia* (Dum.) Dum. (*C. media Lindb.*), *Mannia fragrans* (Balbis) Frye and Clark, *Odontoschisma prostratum* (Sw.) Trev., *Petalophyllum ralfsii* (Wils.) Nees and Gott., *Porella pinnata* L., *Riccia beyrichiana* Hampe *ex* Lehm., *R. hirta* (Aust.) Underw., *Scapania nemorosa* (L.) Dum., and the Texas bottle liverwort (*Sphaerocarpos texanus* Aust.).

Terrestrial species of sandy, upland soils or moist, acidic lowland habitats typically reach their western limits east of the Balcones Escarpment. Terrestrial species tolerant of clayey or limestone-derived soils and epiphytic, bark-inhabiting species may reach Central Texas. Of special interest are those species of peat or bog mosses (genus *Sphagnum*) that have been collected here as recently as 1973 and only recently rediscovered (Jason Singhurst, pers. comm. 2007, 2008). These are *S. affine* Ren. & Card. (*S. imbricatum* Hornsch. *ex* Russ. var. *affine*), *S. lescurii* Sull. (*S. subsecundum* Nees *ex* Sturm. var. *rufescens* (Nees, Hornsch., & Sturm.) Hub.), and *S. palustre* Linnaeus (*S. recurvum* P. Beauv. *sensu latu*). Two of their frequent associates in bogs, fens, mucklands, and peatlands farther east are also found here: the liverwort *Telaranea longifolia* (*T. nematodes* var. *longifolia*) and veilwort (*Pallavicinia lyellii*), often in association with the moss *Isopterygium tenerum* var. *fulvum,* all at their western limit in the southeastern United States in the Ottine Wetlands.

A few, such as the mosses *Amblystegium varium* (Hedw.) Lindb., *Bryum capillare* Hedw., *Funaria americana* Lindb., *Haplocladium microphyllum* (Hedw.) Broth., *Phascum cuspidatum* Hedw., *Physcomitrium pyriforme* (Hedw.) Hampe, *Weissia controversa* Hedw.; the liverworts *Pellia epiphylla* and *Riccia fluitans;* and the yellow hornwort (*Phaeoceros laevis*), are widespread in North America and often cosmopolitan worldwide.

Two species of macroscopic, aquatic algae also occur here: the stoneworts *Chara zeylanica* and *Nitella flexilis.* Both are species with worldwide, cosmopolitan distributions.

Mosses, liverworts, and hornworts are small plants that, like ferns and horsetails, do not produce seeds or flowers. Mosses are easily the most familiar. Most biologists do not consider algae to be plants of any kind but members of a different kingdom of life altogether. The two algae species treated here are large, macroscopic "stoneworts" that may easily be confused with aquatic flowering plants.

Duckweeds and watermeals

Greater duckweed *Spirodela polyrhiza* (C. Linnaeus)
M. Schleiden
Pale duckweed *Lemna valdiviana* R. Philippi
Common watermeal *Wolffia columbiana* G. Karsten
Dotted watermeal *W. brasiliensis* H. Weddell
LEMNACEAE (DUCKWEED FAMILY)
Figs. 7-1, 7-2
Rarely flowering

Field recognition: Greater duckweed: Tiny, floating, annual or perennial aquatic herb overwintering via dense buds (turions) that sink to the pond bottom, to rise again with renewed activity the following spring. Duckweeds are not known to endure drought in this way, however. Plants greatly reduced with a simplified body (thallus), rounded egg-shaped (3–10 mm by 3–6 mm), dark waxy green above, reddish-purple beneath with several conspicuous radiating veins and a tuft of up to twenty capped rootlets hanging below. Proliferating quickly by budding, often forming a solid green cover over the water surface. Flowers and fruits rarely produced. Of quiet, nutrient-rich (eutrophic) waters: ponds, lakes, bayous, and sluggish streams.

Fig. 7-1
Greater duckweed (*Spirodela polyrhiza*) and two tiny watermeals (*Wolffia columbiana*, light green; *W. brasiliensis*, dark green).

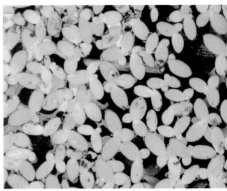

Fig. 7-2
Pale duckweed (*Lemna valdiviana*).

Similar species: The duckweed genus *Lemna* may be recognized by its thalli, each of which possesses a single rootlet rather than a bundle of several. We found pale duckweed (*L. valdiviana*) (Fig. 7-2) nearly covering the surface of several of the "lagoonal" ponds on the San Marcos floodplain in Palmetto State Park.

Field recognition: Watermeals: Tiny, highly reduced, pinhead-sized, rootless aquatic annual or perennial herbs, floating at or just below the surface of the water. Quickly spreading via budding and forming a pale green, granular "scum" several plants thick at or below the water surface. Flowers and fruits rarely produced. Of meso- and eutrophic quiet waters in temperate and subtropical regions worldwide. Mainly the eastern half of the United States. Two species occur here and may be easily distinguished (with magnification) by the presence (dotted watermeal, *W. brasiliensis*) or absence (common watermeal, *W. columbiana*) of a dorsal, nipplelike point on the upper surface.
Similar species: None.
Remarks: Greater duckweed is a small, floating, aquatic plant less than 1 cm in diameter. Though less than a centimeter long, it is the largest of the surface-floating duckweeds (Correll and Johnston 1970), a veritable giant among its confamilials. In the photograph, it is in the company of two tiny watermeal species that happen to be not only smaller than duckweed but the smallest of all flowering plants. About the size of a pinhead, common watermeal is the lighter-colored species; dotted watermeal is the darker. Up to 2 million individuals may occupy a single square meter of pond surface (Diggs, Lipscomb, and O'Kennon 1999). Compare them with the giant sequoia (*Sequoiadendron giganteum* (J. Lindley) J. Buchholz) and the coast redwood (*Sequoia sempervirens* (D. Don) J. Endlicher), two of the world's largest vascular seed plants.

Distribution: Greater duckweed is widely distributed in North America and on other continents. The two watermeals occur mainly in the eastern half of the United States and widely in temperate and subtropical regions worldwide.

MOSSES

Bush's fissidens moss
Fissidens bushii (Card. & Ther.) Card. & Ther.
FAMILY FISSIDENTACEAE
Fig. 7-3
Sporulating winter–spring

Field recognition: Small, dark green plants (gametophytes) with stems less than 1 cm tall. Leaves in two rows, folded in half, and sheathing each other basally, giving the leafy stems a flattened, frondlike appearance. Leaves without a border of elongate cells and with minutely scalloped-saw-toothed margins (use magnification). Stalked spore capsules (sporophytes) arising laterally from the gametophyte "fronds," about 1 cm long on orange, hairlike stalks 6–10 mm long with more or less erect spore capsules. In sandy or clayey soils or sometimes rocks in open woods and the banks of upland, intermittent drainages.

Similar species: *F. bryoides* Hedwig is a species of shady, moister lowland habitats in calcareous or clay soils and sometimes on bark at the bases of tree trunks. It may be distinguished by its terminal sporophytes and leaves with bordered (by a margin of elongate cells 1–3 cells wide), smooth margins.

Distribution: *F. bushii* ranges throughout the eastern United States and Canada through the eastern half of Texas; *F. bryoides* occurs throughout North America, Europe, and Asia to Africa and Australia.

Fig. 7-3
The moss *Fissidens bushii.*

Isopterygium moss
Isopterygium tenerum (Sw.) Mitt.
FAMILY HYPNACEAE
Fig. 7-4
Sporulating in winter

Field recognition: Small plants (gametophytes) in shiny, yellow-green mats of
  irregularly branching stems. Leaves erect-spreading and folded lengthwise
  and flattened (complanate), with twisted, flexuous, long-pointed tips. Spore
  capsules (sporophytes) with orange-brown to reddish hairlike stalks, to 20
  mm long with horizontal to pendulous curved capsules. Of dryish, sandy soils
  and on logs, stumps, and bark at the bases of tree trunks.

Similar species: Variety *fulvum* (Hook. & Wils.) Paris appears to be a more
  robust form growing in moist to wet habitats with larger, darker, often brown-
  ish leaves and dark, reddish stems. It seems to grade into the drier habitat
  form described above. On decaying wood, tree bases, soil and acidic peaty,
  mucky, or sandy banks of swamps, bogs, and seepages.

Remarks: An associate of *Telaranea longifolia*, veilwort (*Pallavicinia lyellii*), and
  perhaps the peat mosses (*Sphagnum* spp.) that have recently been redocumented here (Jason Singhurst, pers. comm. 2007, 2008).

Distribution: Throughout the southeastern United States through East and
  Southeast Texas to the Edwards Plateau.

Fig. 7-4
The moss *Isopterygium
tenerum* with spore
capsules and veilwort
(*Pallavicinia lyelli*).

LIVERWORTS

Telaranea liverwort
Telaranea longifolia (Howe) Engel & Merrill
FAMILY LEPIDOZIACEAE
Fig. 7-5
Sporulating in fall (post-September)

Field recognition: Minute, translucent gametophytes consisting of delicate, irregularly branching, pale green, algalike strands less than 0.5 mm thick with leaves of 2–4 hairlike lobes divided to the base, each 4–8 cells long and 1–2 cells wide. Sporophytes budlike and surrounded by an involucre of divided leaves, rarely seen. Growing intermixed with other bryophytes, especially veilwort and the moss *Isopterygium tenerum* in the Ottine Wetlands, as well as peat mosses (*Sphagnum* spp.), on wet peaty to sandy-peaty substrates.

Similar species: No other liverwort of the area has such a filamentous habit.

Remarks: This tiny, threadlike, leafy liverwort is a bog indicator that often grows intertwined among other acid-loving bryophytes such as veilwort and the mosses *I. tenerum* var. *fulvum* and *Sphagnum* spp. Prior to our record of it in the Ottine Wetlands, the westernmost reports of its occurrence were southeastern Oklahoma (Talbot and Ireland 1982; Studlar and McAlister 1994) and the Big Thicket of Southeast Texas (Stoneburner and Wyatt 1979). It is a very distinctive bryophyte with its pale green, threadlike stems and forked leaves consisting of paired filaments, though it requires magnification to be appreciated.

Distribution: *Telaranea longifolia,* formerly considered a variety of *T. nematodes* (Gott. *ex* Aust.) Howe (as var. *longifolia* Howe), ranges throughout the Atlantic and Gulf Coastal Plain of the eastern United States, from eastern Long Island and Martha's Vineyard south and west to the Ottine Wetlands (Engel and Merrill 2004).

Fig. 7-5
The liverwort
Telaranea longifolia
(center) with veilwort
(*Pallavicinia lyellii*).

Veilwort
Pallavicinia lyellii (Hook.) Gray
PALLAVICINIACEAE (VEILWORT FAMILY)
Fig. 7-6
Sporulating February–April

Field recognition: Gametophytes of pure translucent green, straplike thalli 2–6 cm long and 4–5 mm wide with a thickened midrib containing a dark, threadlike central strand (visible when backlighted) of elongate cells. Branching occasionally and forming flat mats on wet, acidic sandy or peaty banks of seeps, creeks, swamps, and bogs. Sporophytes consisting of black or dark brown ellipsoid capsules borne aloft on pale whitish stalks, splitting into four valves and releasing brown spores with the aid of coiled, springlike elaters that cling like rusty brown wool to the tips and inner surfaces of the valves.

Fig. 7-6
Veilwort (*Pallavicinia lyellii*) with spore capsules.

Similar species: The translucent green, straplike gametophyte of *Pallavicinia* may be distinguished from that of *Pellia epiphylla* (C. Linnaeus) Lindb. (Fig. 7-7) by the dark, threadlike strand of thick-walled cells that runs down its center, which is absent in the latter species.

Remarks: This is a liverwort that grows on sandy, acidic seepage banks and boggy areas. It often occurs intermixed with other acidophiles, such as

Fig. 7-7
The liverwort *Pellia epiphylla.*

Telaranea longifolia and peat or bog mosses (*Sphagnum* spp.). Three species of *Sphagnum* (*S. affine, S. lescurii,* and *S. palustre*) have been collected from the Ottine Wetlands as recently as 1973 (Lodwick and Snider 1980) and represent the southwestern limit of the genus in the eastern United States. We saw no *Sphagnum* mosses during the course of our explorations. However, it has recently been rediscovered here and elsewhere in Gonzales County (Jason Singhurst, pers. comm. 2007, 2008). Perhaps *Pallavicinia* and *Telaranea,* both also at the western limits of their respective ranges in the United States, represent the last indicators of this relict, bog bryophyte community in Central Texas. Sporophytes appear as black, ellipsoid capsules supported by transparent, threadlike setae growing from short green, gametophyte sheaths along the center of the upper surfaces of the thalli.

Distribution: Veilwort ranges throughout the eastern United States from Canada to Florida west to a limit in or near the Ottine Wetlands.

Crystalwort
Riccia fluitans C. Linnaeus
FAMILY RICCIACEAE
Fig. 7-8
Sporulating late spring, summer

Field recognition: Gametophyte body (thallus) pale yellow-green, of thin, narrow, forking strips or ribbons (0.5–0.8 mm wide and 1–5 cm long) with notched tips, forming tangled masses floating just beneath the surface of the water. Floating plants sterile. Plants stranded on mud may produce ventrally bulging spore capsules. Of still or sluggish, quiet waters of ditches, ponds, or their margins.

Similar species: None.

Remarks: Along with *Ricciocarpus natans* (C. Linnaeus) Corda, the only truly aquatic liverworts in North America.

Distribution: Cosmopolitan.

Fig. 7-8
The aquatic liverwort
Riccia fluitans.

HORNWORTS

Yellow hornwort
Phaeoceros laevis (C. Linnaeus) Prosk.
ANTHOCEROTACEAE (HORNWORT FAMILY)
Fig. 7-9
Sporulating February–April

Field recognition: Gametophyte body (thallus) of dark green rosettes to 3 cm across and with irregularly lobed or ruffled (when crowded) edges. Cells of thallus with one or two large, lozenge-shaped chloroplasts. Sporophytes linear, pale yellow-green, sprouting and elongating from their bases like grass from short (1–4 mm), erect, cylindrical involucres of the thallus surface, becoming 2.50–3.75 cm long. Sporophyte splits and blackens from the tip and releases yellow spores at maturity. Grows on wet rocks and moist soil of creek and stream banks.

Similar species: The liverwort *Pellia epiphylla* (C. Linnaeus) Lindb. (Fig. 7-7 [p. 220]) occurs in similar habitats, and its thalloid gametophyte is very similar. It tends to grow in a more directional manner, forming elongate, straplike rather than circular thalli. However, only a microscopic examination of its leaf cells with their numerous, smaller chloroplasts will distinguish the gametophytes of these two species with certainty.

Remarks: This bryophyte grows on wet banks and rocks in Rutledge Swamp and along Rutledge Creek. The circular thalli are a dark blackish-green and often occur in masses. They contain symbiotic colonies of cyanobacteria (*Nostoc* spp.) that may provide them with fixed nitrogen (nitrates from atmospheric nitrogen gas) in the nutrient-poor acidic environments where they grow. The sporophytes are very distinctive, growing like long, thin, yellow-green blades of grass from the dark thalli. At maturity they split at the tips, releasing yellow spores. Sterile thalli may be identified by observing the cells under a

Fig. 7-9
Yellow hornwort
Phaeoceros laevis
with emerging spore
capsules.

microscope. Each will be found to contain a single (or occasionally two) large, lozenge-shaped chloroplast rather than many smaller ovoid ones, as do the liverworts.

Distribution: This species is widespread throughout the Northern Hemisphere.

ALGAE

Stonewort, musk-grass
Chara zeylanica Klein *ex* C. L. Willdenow
CHARACEAE (STONEWORT FAMILY)
Fig. 7-10
Fruiting (oogonia) in summer

Field recognition: Lax, bright green aquatic algae often forming extensive submerged beds attached to the bottom and floating just below the water surface. In calcareous waters often encrusted with lime (calcium carbonate) and having a musky or garlicky odor. Stems jointed and with conspicuous whorls of unbranched branchlets at the nodes. Tiny ovoid oogonia, red to blackish, in rows along upper side of fertile branchlets.

Similar species: See following species (*Nitella flexilis*).

Remarks: We saw great beds of this giant alga in the old fish hatchery pond that feeds the cattail marsh near the southwestern boundary of Palmetto State Park.

Distribution: Worldwide with numerous varieties from tropical to boreal latitudes (Wood and Imahari 1964, 1965).

Fig. 7-10
Musk-grass (*Chara zeylanica*).

Bassweed
Nitella flexilis (C. Linnaeus) C. A. Agardh
CHARACEAE (STONEWORT FAMILY)
Fig. 7-11
Fruiting (oogonia) in summer

Field recognition: Short, densely tufted or clumped aquatic algae with a translucent, dark blue- or blackish-green color. Usually of acidic to pH-neutral waters, odorless and of soft or smooth texture without lime (calcium carbonate) encrustation. Stem internodes contracted and whorled branchlets secondarily branching to produce densely branched glomerules at intervals along the shortened stem axis. Oogonia dark, blackish.

Similar species: The genus *Nitella* may be distinguished from *Chara* by the secondarily branched, rather than unbranched, branchlets arising in whorls from the stem nodes, and its usual lack of a skunky odor.

Remarks: We observed an apparently dwarfed growth form of this species growing in shady seepage water only a few centimeters deep in Palmetto State Park below the old fish hatchery pond a short distance above the cattail/sedge marsh. Its dark green, almost black color and short, tufted habitus was very reminiscent of *Sphagnum* moss, which was our first thought until a closer examination revealed it to be a species of giant algae.

Distribution: Worldwide and with numerous varieties from tropical to boreal latitudes (Wood and Imahari 1964, 1965).

Fig. 7-11
The stonewort *alga Nitella flexilis.*

Appendix 1:
Texas-Endemic Plants of the Ottine Wetlands

1. Texas umbrellawort (*Tauschia texana*): endemic to roughly one dozen counties in southern Texas
2. Wright's false mallow (*Malvastrum aurantiacum*): endemic to Central and southern Texas
3. Sandyland Texas bluebonnet (*Lupinus subcarnosus*): endemic to the southern and southeastern part of the state
4. Texas bluebonnet (*Lupinus texensis*): endemic to the Edwards Plateau and Blackland Prairie regions of the state.
5. Giant spiderwort (*Tradescantia gigantea*): endemic to Central Texas
6. Drummond's phlox (*Phlox drummondii*): endemic to southern Texas until it was introduced elsewhere because of its beauty
7. Drummond's wild-petunia (*Ruellia drummondiana*): endemic from north-central to south-central Texas
8. Texas hawthorn (*Crataegus texana*): endemic to Southeast and coastal Texas
9. Showy buttercup (*Ranunculus macranthus*): endemic to east-central Texas.

Appendix 2:
Alien Plants of the Ottine Wetlands

1. Chinaberry (*Melia azedarach*): native to Asia
2. Umbrella sedge (*Cyperus involucratus*): an Old World native, probably originating in Africa
3. Coffee senna (*Senna occidentalis*): probably a Neotropical native
4. Yellow flag (*Iris pseudacorus*): a native of Europe or Africa
5. Three-lobe false mallow (*Malvastrum coromandelianum*): probably native to tropical America
6. Water-hyacinth (*Eichhornia crassipes*): native to Brazil
7. Dead-nettle (*Lamium amplexicaule*): a European native
8. Milk thistle (*Silybum marianum*): a Mediterranean native
9. Turnsole (*Heliotropium indicum*): a Eurasian native
10. Common dayflower (*Commelina communis*): a native of eastern Asia

Appendix 3:
Previously Published, Untreated Species

(Bogusch 1928, 1930; McAllister, Hogland, and Whitehouse 1930; Tharp 1935; Parks 1935a; Whitehouse and McAllister 1954; Whitehouse 1955; Raun 1958, 1959; Ellison 1964; Lodwick and Snider 1980; Turner et al. 2003a, 2003b)

Liverworts
Aytoniaceae
  Mannia fragrans
Cephaloziaceae
  Cephalozia lunulifolia
  Odontoschisma prostratum
Fossombroniaceae
  Petalophyllum ralfsii
Porellaceae
  Porella pinnata
Ricciaceae
  Riccia beyrichiana
  R. hirta
Scapaniaceae
  Scapania nemorosa
Sphaerocarpaceae
  Sphaerocarpos texanus

Bog, Peat Mosses
Sphagnaceae
  Sphagnum imbricatum
  S. lescurii
  S. palustre

Brown Mosses
Amblystegiaceae
  Amblystegium varium
  Hygroamblystegium tenax
Archidiaceae
  Archidium hallii
Bartramiaceae
  Philonotis longiseta
Ditrichaceae
  Ditrichum pallidum
Entodontaceae
  Entodon seductrix
Funariaceae
  Funaria americana
  Physcomitrium pyriforme
Hypnaceae
  Steerecleus (Rhynchostegium)
    serrulatus
Leskeaceae
  Clasmatodon parvulus
  Haplocladium microphyllum
  Leskea australis
Leucobryaceae
  Leucobryum albidum
Leucodontaceae
  Leucodon julaceus

Polytrichaceae
Atrichum angustatum
Pogonatum brachyphyllum
Pottiaceae
Astomum ludovicianum
A. muhlenbergianum
A. phascoides
Desmatodon plinthobius
Phascum cuspidatum
Weissia controversa

Fern Allies

Selaginellaceae
Riddell's spikemoss (*Selaginella riddellii*)

Ferns

Marsileaceae
Large-footed pepperwort, water-clover (*Marsilea macropoda*)
Hairy pepperwort (*M. vestita*)
Ophioglossaceae
Engelmann's adder's tongue (*Ophioglossum engelmannii*)
Polypodiaceae
Resurrection fern (*Pleopeltis polypodioides* var. *michauxianum*)

Flowering Plants

Dicots

Amaranthaceae
Palmer's amaranth (*Amaranthus palmeri*)
Anacardiaceae
Winged sumac (*Rhus copallina*)
Skunk bush (*R. trilobata*)
Apiaceae
Hoary bowlesia (*Bowlesia incana*)
Wild-celery (*Ciclospermum leptophyllum*)
Southern carrot (*Daucus pusillus*)
Floating water-pennywort (*Hydrocotyle ranunculoides*)

Forked scaleseed (*Spermolepis divaricata*)
Apocynaceae
Texas blue-star (*Amsonia ciliata* var. *texana*)
Aristolochiaceae
Swan flower (*Aristolochia erecta*)
Asclepiadaceae
Blunt-leaf milkweed (*Asclepias amplexicaulis*)
Antelope horns (*A. asperula* var. *capricornu*)
Swamp milkweed (*A. incarnata*)
Hierba de zizotes (*A. oenotheroides*)
Swallow-wort (*Cynanchum laeve*)
Tayalote (*C. unifarium*)
Short-crowned milkvine (*Matelea brevicoronata*)
Common milkvine (*M. cynanchoides*)
Asteraceae
Short ragweed (*Ambrosia artemisiifolia*)
Western ragweed (*A. psilostachya*)
Giant ragweed (*A. trifida*)
Arkansas lazy-daisy (*Aphanostephus skirrhobasis*)
Western mugwort (*Artemisia ludoviciana*)
Azurea aster (*Aster azureus*)
Tall aster (*A. praealtus*)
Mexican devil-weed (*A. spinosus*)
Spreading aster (*A. subulatus*)
Soft green-eyes (*Berlandiera pumila*)
Beggar's ticks (*Bidens frondosa*)
Smooth bidens (*B. laevis*)
Common least-daisy (*Chaetopappa asteroides*)
Swamp thistle (*Cirsium muticum*)
Texas thistle (*C. texanum*)
Horseweed (*Conyza canadensis*)
Golden-mane coreopsis (*Coreopsis basalis*)

Swamp coreopsis (*C. cardaminaefolia*)
Crowned coreopsis (*C. nuecensis*)
Scratch-daisy (*Croptilon divaricatum*)
Pieplant (*Eclipta alba*)
Engelmann-daisy (*Engelmannia pinnatifida*)
American burnweed (*Erechtites hieracifolia*)
Philadelphia fleabane (*Erigeron philadelphicus*)
Dog-fennel (*Eupatorium capillifolium*)
Yankee weed (*E. compositifolium*)
Pink boneset (*E. incarnatum*)
Fall boneset (*E. serotinum*)
Silver cotton-rose (*Evax candida*)
Indian blanket (*Gaillardia pulchella*)
Sweet gaillardia (*G. suavis*)
Falcate cudweed (*Gamochaeta falcata*)
Purple cudweed (*G. purpurea*)
Tarweed (*Grindelia squarrosa*)
Texas broomweed (*Gutierrezia texana*)
Bitterweed (*Helenium amarum*)
Small-head sneezeweed (*H. microcephalum*)
Swamp sunflower (*Helianthus angustifolius*)
Silverleaf sunflower (*H. argophyllus*)
Cucumber-leaf sunflower (*H. debilis* var. *cucumerifolia*)
Sawtooth sunflower (*H. grosseserratus*)
Camphor weed (*Heterotheca subaxillaris*)
Flat-top woolly-white (*Hymenopappus scabiosaeus*)
Narrow-leaf sumpweed (*Iva angustifolia*)
Marsh-elder (*I. annua*)

Weedy dwarf-dandelion (*Krigia caespitosa*)
Tall gayfeather (*Liatris aspera*)
Handsome blazing-star (*L. elegans*)
Lindheimer-daisy (*Lindheimera texana*)
Skeleton plant (*Lygodesmia texana*)
Hoary blackfoot-daisy (*Melampodium cinereum*)
Showy palafoxia (*Palafoxia hookeriana*)
False ragweed (*Parthenium hysterophorus*)
Camphor weed (*Pluchea camphorata*)
Tuberous false dandelion (*Pyrrhopappus grandiflorus*)
Yellow sanvitalia (*Sanvitalia ocymoides*)
Mexican bone bract (*Sclerocarpus uniserialis*)
Butterweed (*Senecio glabellus*)
Golden groundsel (*S. obovatus*)
Groundsel (*S. tampicanus*)
Bear's foot (*Smallanthus uvedalia*)
Boott's goldenrod (*Solidago arguta* var. *boottii*)
Giant goldenrod (*S. gigantea*)
Prickly sow-thistle (*Sonchus asper*)
Common sow-thistle (*S. oleraceus*)
Cowpen-daisy (*Verbesina encelioides*)
Guadalupe ironweed (*Vernonia* X *guadalupensis* (*V. baldwinii* X *V. lindheimeri*))
Plateau goldeneye (*Viguiera dentata*)

Boraginaceae
Hairy stickseed (*Lappula occidentalis*)
Carolina puccoon (*Lithospermum caroliniense*)
Narrow-leaf puccoon (*L. incisum*)

Brassicaceae
Sicklepod (*Arabis canadensis*)

Shepherd's purse (*Capsella bursa-
pastoris*)
Sand bittercress (*Cardamine
parviflora*)
Bittercress (*C. pensylvanica*)
Short-pod draba (*Draba
brachycarpa*)
Wedgeleaf draba (*D. cuneifolia*)
Big-flower bladderpod (*Lesquerella
grandiflora*)
Slender bladderpod (*L. recurvata*)
Watercress (*Rorippa nasturtium-
aquaticum*)
White mustard (*Sinapsis alba*)
Hedge-mustard (*Sisymbrium
officinale*)
Cactaceae
Cory cactus (*Coryphantha
missouriensis*)
Lace cactus (*Echinocereus
riechenbachii*)
Texas prickly-pear (*Opuntia
lindheimeri*)
Callitrichaceae
Large water-starwort (*Callitriche
heterophylla*)
Campanulaceae
Small Venus' looking-glass
(*Triodanus biflora*)
Clasping Venus' looking-glass (*T.
perfoliata*)
Capparaceae
Narrow-leaf rhombopod (*Cleomella
angustifolia*)
Large clammy weed (*Polanisia
erosa*)
Caprifoliaceae
White honeysuckle (*Lonicera
albiflora*)
Possumhaw viburnum (*Viburnum
nudum*)
Caryophyllaceae
Drummond's nailwort (*Paronychia
drummondii*)

Chenopodiaceae
Pale goosefoot (*Chenopodium
albescens*)
Common lamb's quarters (*C.
album*)
Epazote (*C. ambrosioides*)
Convolvulaceae
Heart-leaf morning glory (*Ipomoea
cordatotriloba*)
Saltmarsh morning glory (*I.
sagitta*)
Cucurbitaceae
Buffalo gourd (*Cucurbita
foetidissima*)
Lindheimer's globe berry (*Ibervillea
lindheimeri*)
Cuscutaceae
Cluster dodder (*Cuscuta glomerata*)
Ericaceae
Staggerbush (*Lyonia mariana*)
Farkleberry (*Vaccinium arboreum*)
Euphorbiaceae
Mala mujer (*Cnidoscolus texanus*)
Woolly croton, goatweed (*Croton
capitatus*)
Tropic croton (*C. glandulosus*)
One-seed croton (*C.
monanthogynous*)
Texas croton (*C. texensis*)
Snow-on-the-prairie (*Euphorbia
(Agaloma) bicolor*)
White-margined spurge (*E.
(Chamaesyce) albomarginata*)
Wild poinsettia (*E. (Poinsettia)
cyathophora*)
Toothed spurge (*E. (P.) dentata*)
Warty spurge (*E. (Tithymalus)
spathulata*)
Castor-bean (*Ricinus communis*)
Queen's delight (*Stillingia sylvatica*)
Noseburn (*Tragia ramosa*)
Fabaceae
Huisache (*Acacia minuata*)
Southern hog-peanut
(*Amphicarpaea bracteata*)

Berlandier's ground-plum
(*Astragalus crassicarpus* var.
berlandieri)
Nuttall's milk-vetch (*A.*
nuttallianus)
Round-fruit wild-indigo (*Baptisia*
sphaerocarpa)
Sicklepod senna (*Cassia obtusifolia*)
Butterfly-pea (*Centrosema*
virginianum)
Pigeon-wings (*Clitoria mariana*)
Wedgeleaf prairie-clover (*Dalea*
emarginata)
Long-bract prairie-clover (*D.*
phleoides var. *microphylla*)
Slender prairie-clover (*D. tenuis*)
Hoary tick-clover (*Desmodium*
canescens)
Smooth tick-clover (*D. glabellum*)
Panicled tick-clover (*D.*
paniculatum)
Honey locust (*Gleditsia*
triancanthos)
Scarlet-pea (*Indigofera miniata*)
Indigo (*I. suffruticosa*)
Low pea-vine (*Lathyrus pusillus*)
Trailing bush-clover (*Lespedeza*
procumbens)
Creeping bush-clover (*L. repens*)
Violet bush-clover (*L. violacea*)
Spotted bur-clover (*Medicago*
arabica)
Alfalfa, lucerne (*M. sativa*)
White sweet-clover (*Melilotus*
albus)
American snout-bean (*Rhynchosia*
americana)
Coffee-bean (*Sesbania macrocarpa*)
Texas mountain-laurel (*Sophora*
secundiflora)
Lindheimer's hoary-pea (*Tephrosia*
lindheimeri)
Multi-bloom hoary-pea (*T.*
onobrychoides)
Devil's shoestring (*T. virginiana*)

Bejar clover (*Trifolium bejariense*)
Carolina clover (*T. carolinianum*)
Peanut clover (*T. polymorphum*)
Red clover (*T. pratense*)
Leavenworth's vetch (*Vicia*
leavenworthii)
Pygmy-flowered vetch (*V.*
minutiflora)
Viperina (*Zornia bracteata*)
Fagaceae
White oak (*Quercus alba*)
Texas oak (*Q. buckleyi* (*texana*))
Southern red oak (*Q. falcata*)
Blackjack oak (*Q. marilandica*)
Post oak (*Q. stellata*)
Live oak (*Q. virginiana*)
Fumariaceae
Golden corydalis (*Corydalis aurea*)
Curved-pod corydalis (*C.*
curvisiliqua)
Geraniaceae
Carolina geranium (*Geranium*
carolinianum)
Haloragaceae
Parrot's feather (*Myriophyllum*
heterophyllum)
Hydrophyllaceae
Rough nama (*Nama hispidum*)
Fiddle-leaf nama (*N. jamaicense*)
Small-flowered nemophila
(*Nemophila aphylla*)
Woolly blue-curls (*Phacelia*
congesta)
Hairy phacelia (*P. hirsuta*)
Prairie phacelia (*P. strictiflora*)
Hypericaceae
St. Andrew's cross (*Ascyrum*
hypericoides)
Clasping St. John's wort (*Hypericum*
mutilum)
Juglandaceae
Mockernut hickory (*Carya alba*)
Black hickory (*C. texana*)
Black walnut (*Juglans nigra*)

Lamiaceae
  Blunt-sepal brazoria (*Brazoria truncata*)
  Wild bergamot (*Monarda fistulosa*)
  Intermediate lion's-heart (*Physostegia intermedia*)
  Cedar sage (*Salvia roemeriana*)
  Drummond's skullcap (*Scutellaria drummondii*)
  Shade betony (*Stachys crenata*)
  Wood germander (*Teucrium canadense*)
Loasaceae
  Yellow rock-nettle (*Eucnide bartonioides*)
  Bractless mentzelia (*Mentzelia nuda*)
Lythraceae
  Lance-leaf loosestrife (*Lythrum alatum* var. *lanceolatum*)
Malvaceae
  Fringed poppy-mallow (*Callirhoe digitata*)
  Tall poppy-mallow (*C. leiocarpa*)
  Common mallow (*Malva rotundifolia*)
  Bracted sida (*Sida ciliaris*)
  Spreading sida (*S. filicaulis*)
  Lindheimer's sida (*S. lindheimeri*)
Molluginaceae
  Indian chickweed, carpetweed (*Mollugo verticillata*)
Nyctaginaceae
  Scarlet spiderling (*Boerhavia diffusa*)
Oleaceae
  Mexican ash (*Fraxinus berlandieriana*)
Onagraceae
  Drummond's gaura (*Gaura drummondii*)
  Lizard-tail gaura (*G. parviflora*)
  Roadside gaura (*G. suffulta*)
  Bushy seed-box (*Ludwigia alternifolia*)
  Upright primrose-willow (*L. decurrens*)
  Cylindric-fruited ludwigia (*L. glandulosa*)
  Angle-stem water-primrose (*L. leptocarpa*)
  Shrubby water-primrose (*L. octovalvis*)
  Marsh purslane (*L. palustris*)
  Creeping water-primrose (*L. peploides*)
Oxalidaceae
  Yellow wood-sorrel (*Oxalis dillenii* (*stricta*))
  Yellow wood-sorrel (*O. lyonii*)
Passifloraceae
  Yellow passion flower (*Passiflora lutea*)
Plantaginaceae
  Bottlebrush plantain (*Plantago aristata*)
  Many-seed plantain (*P. heterophylla*)
  Tallow weed (*P. hookeriana*)
Polemoniaceae
  Big-sepal phlox (*Phlox pilosa* subsp. *latisepala*)
Polygalaceae
  White milkwort (*Polygala alba*)
Polygonaceae
  Long-leaf wild-buckwheat (*Eriogonum longifolium*)
  Heart-sepal wild-buckwheat (*E. multiflorum*)
  Pennsylvania smartweed (*Polygonum pensylvanicum*)
  Tearthumb, arrow vine (*P. sagittatum*)
  Hairy smartweed (*P. setaceum*)
  Sheep sorrel (*Rumex acetosella*)
  Pale dock (*R. altissimus*)
  Amamastla (*R. chrysocarpus*)
  Curly dock (*R. crispus*)

Primulaceae
Thin-leaf brookweed (*Samolus parviflorus*)
Ranunculaceae
Texas virgin's bower (*Clematis drummondii*)
Scarlet clematis (*C. texensis*)
Blue larkspur (*Delphinium carolinianum*)
Rhamnaceae
Carolina buckthorn (*Rhamnus caroliniana*)
Lotebush (*Ziziphus obtusifolia*)
Rosaceae
Pasture haw (*Crataegus spathulata*)
Sutherland's hawthorn (*C. sutherlandensis*)
Texas hawthorn (*C. texana*)
Green hawthorn (*C. viridis*)
Mexican plum (*Prunus mexicana*)
Texas almond (*P. texana*)
Rubiaceae
Rough buttonweed (*Diodia teres*)
Virginia buttonweed (*D. virginiana*)
Catchweed bedstraw (*Galium aparine*)
Hairy bedstraw (*G. pilosum*)
Dye bedstraw (*G. tinctorium*)
Southwestern bedstraw (*G. virgatum*)
Small bluets (*Hedyotis crassifolia*)
Greenman's bluets (*H. greenmannii* (*H. parviflora*))
Rutaceae
Hoptree (*Ptelea trifoliata*)
Hercules'-club (*Zanthoxylum clava-herculis*)
Sapindaceae
Mexican buckeye (*Ungnadia speciosa*)
Saururaceae
Lizard's tail (*Saururus cernuus*)
Scrophulariaceae
Lindheimer's paintbrush (*Castilleja purpurea* var. *lindheimeri*)
Prairie paintbrush (*C. purpurea* var. *purpurea*)
Virginia hedge-hyssop (*Gratiola virginiana*)
Toad-flax (*Linaria* (*Nuttallanthus*) *canadensis*)
False foxglove (*Penstemon cobaea*)
Necklace weed (*Veronica peregrina*)
Solanaceae
Chile piquin (*Capsicum annuum* var. *aviculare*)
Ground-cherry (*Physalis cinerascens*)
Clammy ground-cherry (*P. heterophylla*)
Field ground-cherry (*P. mollis*)
Winged ground-cherry (*P. viscosa* var. *spathulaefolia*)
American nightshade (*Solanum americanum*)
Tiliaceae
Carolina basswood (*Tilia americana* var. *caroliniana*)
Ulmaceae
Lindheimer's hackberry (*Celtis lindheimeri*)
Urticaceae
Blunt pellitory (*Parietaria obtusa*)
Pennsylvania pellitory (*P. pensylvanica*)
Low spring nettle (*Urtica chamaedryoides*)
Valerianaceae
Hairy corn-salad (*Valerianella amarella*)
Beaked corn-salad (*V. radiata*)
Narrow-cell corn-salad (*V. stenocarpa*)
Verbenaceae
West Indian lantana (*Lantana camara*)
Desert lantana (*L. macropoda*)
Wild verbena (*Verbena* (*Glandularia*) *bipinnatifida*)
Texas vervain (*Verbena halei*)

Fan-leaf vervain (*V. plicata*)
Pink verbena (*V. pumila*)
Nettle-leaf vervain (*V. urticifolia*)
Violaceae
Primrose-leaf violet (*Viola primulifolia*)
Vitaceae
Pepper vine (*Ampelopsis arborea*)
Raccoon-grape (*A. cordata*)
Cow itch (*Cissus incisa*)
Seven leaf creeper (*Parthenocissus heptaphylla*)
Winter grape (*Vitis vulpina*)

Monocots
Agavaceae
Spotted American-aloe (*Manfreda maculosa*)
Arkansas yucca (*Yucca arkansana*)
Spanish dagger (*Y. treculeana*)
Alismataceae
Long-lobed arrowhead (*Sagittaria montevidensis*)
Burmanniaceae
Cap burmannia (*Burmannia capitata*)
Commelinaceae
Spreading dayflower (*Commelina diffusa*)
Cyperaceae (Sedges)
Amphibious sedge (*Carex amphibola*)
Cedar sedge (*C. planostachys*)
Kidney sedge (*C. reniformis*)
Taper-leaf flat sedge (*Cyperus acuminatus*)
Baldwin flat sedge (*C. croceus*)
Sheathed flat sedge (*C. haspan*)
Branched flat sedge (*C. polystachyos*)
Pine barren flat sedge (*C. retrorsus*)
False nut-grass (*C. strigosus*)
One-flowered flat sedge (*C. uniflorus*)

Jointed spike sedge (*Eleocharis geniculata*)
Septate spike sedge (*E. interstincta*)
Slender spike sedge (*E. tenuis*)
Twisted spike sedge (*E. tortilis*)
Hairy fimbristylis (*Fimbristylis puberula*)
Vahl's fimbristylis (*F. vahlii*)
Western umbrella sedge (*Fuirena simplex*)
Common hemicarpha (*Lipocarpha micrantha*)
White-top umbrella-grass (*Rhynchospora colorata*)
Woolly-grass bulrush (*Scirpus cyperinus*)
Iridaceae
Prairie celestial-lily (*Nemastylis geminiflora*)
Bermuda blue-eyed-grass (*Sisyrinchium angustifolium*)
Juncaceae (Rushes)
Slimpod rush (*Juncus diffusissimus*)
Grass-leaf rush (*J. marginatus*)
Path rush (*J. tenuis*)
Liliaceae
Blue funnel-lily (*Androstephium coeruleum*)
Drummond's sabadilla (*Schoenocaulon drummondii*)
Marantaceae
Powdery thalia (*Thalia dealbata*)
Orchidaceae
Giant helleborine (*Epipactis gigantea*)
Water spider orchid (*Habenaria repens*)
Rose pogonia (*Pogonia ophioglossoides*)
Poaceae (Grasses)
Carolina foxtail (*Alopecurus carolinianus*)
Bushy bluestem (*Andropogon glomeratus*)
Broomsedge (*A. virginicus*)

Carpet grass (*Axonopus fissifolius*)
Bermuda grass (*Cynodon dactylon*)
Jungle-rice (*Echinochloa colona*)
Barnyard grass (*E. crus-galli*)
Canada wild-rye (*Elymus canadensis*)
Teal love grass (*Eragrostis hypnoides*)
Spreading love grass (*E. pectinacea*)
Little barley (*Hordeum pusillum*)
Rice cutgrass (*Leersia oryzoides*)
Ozark grass (*Limnodea arkansana*)
Texas winter grass (*Nassella leucotricha*)
Beaked panic grass (*Panicum anceps*)
Paired rosette grass (*P. (Dichanthelium) dichotomum*)
Few-flowered rosette grass (*P. (D.) oligosanthes*)
Velvet rosette grass (*P. (D.) scoparium*)
Savannah panic grass (*P. gymnocarpon*)
Knot grass (*Paspalum distichum*)
Florida paspalum (*P. floridanum*)
Timothy canary grass (*Phalaris angusta*)

Annual bluegrass (*Poa annua*)
Sugarcane plume grass (*Saccharum giganteum*)
Knot-root bristle grass (*Setaria parviflora*)
Green foxtail (*S. viridis*)
Prairie wedgescale (*Sphenopholis obtusata*)
Dropseed (*Sporobolus* sp.)
Fringed signal grass (*Urochloa ciliatissima*)
Potamogetonaceae
 Shining pondweed (*Potamogeton illinoensis*)
Smilacaceae
 Fiddle-leaf greenbrier (*Smilax bona-nox*)
 Catbrier (*S. glauca*)
 Kidney-leaf greenbrier (*S. renifolia*)
Typhaceae
 Broad-leaf cattail (*Typha latifolia*)
Xyridaceae
 Carolina yellow-eyed-grass (*Xyris caroliniana*)
 Iris-leaf yellow-eyed-grass (*X. laxifolia* var. *iridifolia*)

Appendix 4:
Previously Unpublished Species

(Williams and Watson 1978; Fleenor and Taber, fieldwork 1999–2004)

Giant Algae
Stonewort (*Chara zeylanica*)
Bassweed (*Nitella flexilis*)

Liverworts
Lepidoziaceae
Telaranea nematodes
Pelliaceae
Pellia epiphylla

Brown Mosses
Bryaceae
Bryum capillare
Fissidentaceae
Fissidens bushii
F. bryoides
Pottiaceae
Barbula cancellata

Ferns
Pteridaceae
Southern maidenhair (*Adiantum capillus-veneris*)
Alabama lip-fern (*Cheilanthes alabamensis*)

Conifers
Cupressaceae
Eastern red-cedar (*Juniperus virginiana*)
Taxodiaceae
Bald-cypress (*Taxodium distichum*)

Flowering Plants
Dicots
Acanthaceae
Low wild-petunia (*Ruellia humilis*)
Wild-petunia (*R. nudiflora*)
Aizoaceae
Desert horse-purslane (*Trianthema portulacastrum*)
Amaranthaceae
Slender snake-cotton (*Froelichia gracilis*)
Anacardiaceae
Smooth sumac (*Rhus glabra*)
Apiaceae
Poison hemlock (*Conium maculatum*)
Bushy eryngo (*Eryngium diffusum*)
Nuttall's mock bishop's weed (*Ptilimnium nuttallii*)
Water-parsnip (*Sium suave*)

Naked scaleseed (*Spermolepis
inermis*)
Field hedge-parsley (*Torilis arvensis*)
Knotted hedge-parsley (*T. nodosa*)
Apocynaceae
Indian-hemp (*Apocynum
cannabinum*)
Asclepiadaceae
Butterfly-weed (*Asclepias tuberosa*)
Anglepod (*Matelea gonocarpos*)
Asteraceae
Texas aster (*Aster drummondii* var.
texanus)
Western daisy (*Astranthium
integrifolium*)
Malta star-thistle (*Centaurea
melitensis*)
Plains coreopsis (*Coreopsis
tinctoria*)
White-top, prairie fleabane
(*Erigeron strigosus*)
Prairie gaillardia (*Gaillardia
fastigiata*)
Purple cudweed (*Gamochaeta
purpurea*)
Common sunflower (*Helianthus
annuus*)
Wild lettuce (*Lactuca canadensis*)
Florida lettuce (*L. floridana*)
Small palafoxia (*Palafoxia callosa*)
Carolina false dandelion
(*Pyrrhopappus carolinianus*)
Mexican hat (*Ratibida columnifera*)
Black-eyed Susan (*Rudbeckia hirta*)
Awnless bush-sunflower (*Simsia
calva*)
Boraginaceae
Spring forget-me-not (*Myosotis
macrosperma*)
Brassicaceae
Pinnate tansy-mustard
(*Descurainia pinnata*)
Hedge-mustard (*Sisymbrium
officinale*)

Caryophyllaceae
Four-leaf many-seed (*Polycarpon
tetraphyllum*)
Cistaceae
Carolina sunrose (*Helianthemum
carolinianum*)
Euphorbiaceae
Lindheimer's copperleaf (*Acalypha
lindheimeri*)
Fabaceae
Plains wild-indigo (*Baptisia
leucophaea*)
Eastern redbud (*Cercis canadensis*
var. *canadensis*)
Texas bluebonnet (*Lupinus texensis*)
Small bur-clover (*Medicago
minima*)
Annual yellow sweet-clover
(*Melilotus indicus*)
Yellow sweet-clover (*M. officinalis*)
Yellow puff (*Neptunia lutea*)
Retama (*Parkinsonia aculeata*)
Round-leaf scurf-pea (*Pediomelum
rhombifolium*)
Sensitive-brier (*Schrankia uncinata*)
Eve's necklace (*Sophora affinis*)
Fagaceae
Water oak (*Quercus nigra*)
Haloragaceae
Brazilian parrot's feather
(*Myriophyllum brasiliense*)
Lamiaceae
Common horehound (*Marrubium
vulgare*)
Lemon beebalm (*Monarda
citriodora*)
Spotted beebalm (*M. punctata*)
Loganiaceae
Florida pinkroot (Texas pinkroot
in part) (*Spigelia loganioides* (*S.
texana* in part))
Malvaceae
Wine cup (*Callirhoe involucrata*)
Yellow false mallow (*Malvastrum
aurantiacum*)

Three-lobe false mallow (*M.
coromandelianum*)
Carolina modiola (*Modiola
caroliniana*)
Arrow leaf sida (*Sida rhombifolia*)
Moraceae
Bois d'arc, Osage-orange (*Maclura
pomifera*)
Oxalidaceae
Pink wood-sorrel (*Oxalis violacea*)
Papaveraceae
White prickly poppy (*Argemone
albiflora*)
Passifloraceae
Maypop (*Passiflora incarnata*)
Polygonaceae
Bushy smartweed (*Polygonum
ramossimum*)
Annual dock (*Rumex hastatulus*)
Fiddle dock (*R. pulcher*)
Rosaceae
Parsley hawthorn (*Crataegus
marshallii*)
Rubiaceae
Star-violet, prairie bluets (*Hedyotis
nigricans*)
One-flowered bluets (*Oldenlandia
uniflora*)
Scrophulariaceae
Texas paintbrush (*Castilleja
indivisa*)
Yellow water-hyssop (*Mecardonia
procumbens*)
Solanaceae
Virginia ground-cherry (*Physalis
virginiana* var. *sonorae*)
Western horse-nettle (*Solanum
dimidiatum*)
Silver-leaf nightshade (*S.
eleagnifolium*)
Sterculiaceae
Anglepod melochia (*Melochia
pyramidata*)

Urticaceae
Florida pellitory (*Parietaria
floridana*)
Verbenaceae
Wedgeleaf frog-fruit (*Phyla
cuneifolia*)
Rough-leaf vervain (*Verbena scabra*)
Vitaceae
Sweet grape (*Vitis cinerea*)
Mustang grape (*V. mustangensis*)
Zygophyllaceae
Puncture vine (*Tribulus terrestris*)

Monocots
Alismataceae
Burhead (*Echinodorus berteroi*)
Arecaceae
Texas, Mexican palmetto (*Sabal
mexicana*)
Cyperaceae (Sedges)
Oval-leaf sedge (*Carex
cephalophora*)
Cherokee sedge (*C. cherokeensis*)
Thin-scale sedge (*C. hyalinolepis*)
Reflexed sedge (*C. retroflexa*)
Tropical flat sedge (*C. surinamensis*)
Jointed flat sedge, chintul (*Cyperus
articulatus*)
Large-spike spike sedge (*Eleocharis
palustris*)
Iridaceae
Purple pleat-leaf (*Alophia
drummondii*)
Herbertia (*Herbertia lahue*)
Yellow flag (*Iris pseudacorus*)
Sword-leaf blue-eyed-grass
(*Sisyrinchium chilense*)
Juncaceae (Rushes)
Inland rush (*Juncus interior*)
Roundhead rush (*J. validus*)
Lemnaceae
Duckweed (*Lemna valdiviana*)
Common duckmeat (*Spirodela
polyrhiza*)

Dotted watermeal (*Wolffia brasiliensis*)
Common watermeal (*W. columbiana*)
Liliaceae
Giant rain-lily (*Cooperia pedunculata*)
Carolina spider-lily (*Hymenocallis caroliniana*)
Poaceae (Grasses)
Purple threeawn (*Aristida purpurea*)
Giant reed (*Arundo donax*)
Silver bluestem (*Bothriochloa laguroides*)
Rescue grass (*Bromus catharticus*)
Japanese brome (*B. japonicus*)
Buffalo grass (*Buchloe dactyloides*)
Sandbur (*Cenchrus spinifex*)
Hooded windmill grass (*Chloris cucullata*)
Hairy crab grass (*Digitaria sanguinalis*)
Virginia wild-rye (*Elymus virginicus*)

Plains love grass (*Eragrostis intermedia*)
Mourning love grass (*E. lugens*)
Red love grass (*E. secundiflora* subsp. *oxylepis*)
Perennial ryegrass (*Lolium perenne*)
Satin grass (*Muhlenbergia schreberi*)
Lindheimer's rosette grass (*Panicum (Dicanthelium) acuminatum* var. *lindheimeri*)
Rosette grass (*P. (D.) nodatum*)
Reed canary grass (*Phalaris canariensis*)
Rabbit's-foot grass (*Polypogon monspeliensis*)
Johnson grass (*Sorghum halapense*)
Pontederiaceae
Water-hyacinth (*Eichhornia crassipes*)
Potamogetonaceae
Threadleaf pondweed (*Potamogeton pusillus*)
Typhaceae
Southern tule (*Typha domingensis*)

Appendix 5:
Complete Checklist of Plants

* = records previously published in local treatments, not treated in main text (Bogusch 1928, 1930; McAllister, Hoglund, and Whitehouse 1930; Tharp 1935; Parks 1935a; Whitehouse and McAllister 1954; Whitehouse 1955; Raun 1958, 1959; Lodwick and Snider 1980; Turner et al. 2003a, 2003b).

** = records previously unpublished in local treatments (Williams and Watson 1978; Fleenor and Taber, fieldwork 1999–2004)

*Acacia minuta (Huisache)
**Acalypha lindheimeri (Lindheimer's copperleaf)
Acer negundo (Boxelder)
**Adiantum capillus-veneris (Southern maidenhair fern)
Aesculus pavia (Red buckeye)
Agalinis purpurea (Purple gerardia)
Allium canadense (Wild onion)
A. drummondii (Drummond's onion)
*Alopecurus carolinianus (Carolina foxtail)
**Alophia drummondii (Purple pleat-leaf)
**Amblystegium varium (Moss)
Amorpha fruticosa (Bastard indigo)
Ampelopsis arborea (Pepper vine)
A. cordata (Raccoon-grape)
*Amphicarpaea bracteata (Southern hog-peanut)
*Amsonia ciliata var. *texana* (Texas blue-star)
A. tabernaemontana (Willow slimpod)

*Andropogon glomeratus (Bushy bluestem)
*A. virginicus (Broomsedge)
*Androstephium coeruleum (Blue funnel-lily)
Anemone berlandieri (Ten-petal thimbleweed)
**Apocynum cannabinum (Indian-hemp)
*Arabis canadensis (Sicklepod)
*Archidium hallii (Moss)
Arisaema dracontium (Green dragon)
**Aristida purpurea (Purple threeawn)
*Aristolochia erecta (Swan flower)
*Artemisia ludoviciana (Western mugwort)
**Arundo donax (Giant reed)
*Asclepias amplexicaulis (Blunt-leaf milkweed)
*A. incarnata (Swamp milkweed)
*A. oenotheroides (Hierba de zizotes)
*Aster azureus (*A. oolentangiensis*) (Azure aster)

***A. drummondii* var. *texanus* (Texas aster)

A. ericoides (White prairie aster)

**A. praealtus* (Tall aster)

**Astomum ludovicianus* (Moss)

**A. muhlenbergianum* (Moss)

**A. phascoides* (Moss)

**Astragalus crassicarpus* var. *berlandieri* (Berlandier's ground-plum)

***Astranthium integrifolium* (Western-daisy)

**Axonopus fissifolius* (Carpet grass)

Azolla caroliniana (Mosquito fern)

Baccharis neglecta (Roosevelt weed)

Bacopa monnieri (Coastal water-hyssop)

**Baptisia sphaerocarpa* (Round-fruit wild-indigo)

***Barbula cancellata* (Moss)

Berchemia scandens (Alabama supplejack)

***Berlandiera pumila* (Soft green-eyes)

*Bidens laevis (Smooth bidens)

Boehmeria cylindrica (Bog-hemp)

**Boerhavia diffusa* (Scarlet spiderling)

***Bothriochloa laguroides* (Silver bluestem)

*Bowlesia incana (Hoary bowlesia)

***Bromus catharticus* (Rescue grass)

***B. japonicus* (Japanese brome)

**Bryum capillare (Moss)

***Buchloe dactyloides* (Buffalo grass)

**Burmannia capitata* (Cap burmannia)

Callicarpa americana (American beautyberry)

**Callirhoe leiocarpa* (Tall poppy-mallow)

**Callitriche heterophylla* (Large water-starwort)

Calyptocarpus vialis (Prostrate-sunflower)

Campsis radicans (Trumpet-creeper)

*Capsicum annuum var. *aviculare* (Chile piquin)

**Cardamine parviflora* (Sand bittercress)

**C. pensylvanica* (Bittercress)

Cardiospermum halicacabum (Balloon vine)

*Carex amphibola (Amphibious sedge)

***C. cephalophora* (Oval-leaf sedge)

***C. cherokeensis* (Cherokee sedge)

C. crus-corvi (Crowfoot caric sedge)

***C. hyalinolepis* (Thin-scale sedge)

C. longii (Bayard Long caric sedge)

C. lupulina (Hop caric sedge)

C. lurida (Sallow caric sedge)

C. microdonta (Small-tooth caric sedge)

**C. planostachys* (Cedar sedge)

**C. reniformis* (Kidney sedge)

***C. retroflexa* (Reflexed sedge)

*Carya alba (Mockernut hickory)

C. illinoinensis (Pecan)

**Cassia obtusifolia* (Sicklepod senna)

***Castilleja indivisa* (Texas paintbrush)

**C. purpurea* var. *lindheimeri* (Lindheimer's paintbrush)

**C. purpurea* var. *purpurea* (Prairie paintbrush)

Celtis laevigata (Southern hackberry)

**C. lindheimeri* (Lindheimer's hackberry)

***Cenchrus spinifex* (Sandbur)

***Centaurea melitensis* (Malta star-thistle)

Cephalanthus occidentalis (Buttonbush)

Chaerophyllum tainturieri (Chervil)

**Chara zeylanica (Stonewort, a green alga)

Chasmanthium latifolium (Wood-oats)

**Chenopodium albescens* (Pale goosefoot)

***Chloris cucullata* (Hooded windmill grass)

**Ciclospermum leptophyllum* (Wild-celery)

Cicuta maculata (Water-hemlock)
***Cirsium horridulum* (Bull thistle)
**C. muticum* (Swamp thistle)
Clematis pitcheri (Leatherflower)
**C. texensis* (Scarlet clematis)
**Cleomella angustifolia* (Narrow-leaf
rhombopod)
*Clitoria mariana (Pigeon-wings)
***Cnidoscolus texanus* (Mala mujer)
Cocculus carolinus (Carolina
snailseed)
Commelina communis (Common
dayflower)
**C. diffusa* (Spreading dayflower)
C. erecta (Erect dayflower)
**Conium maculatum (Poison-
hemlock)
Cooperia drummondii (Drummond's
rain-lily)
***C. pedunculata* (Giant rain-lily)
**Coreopsis nuecensis* (Crowned
coreopsis)
***C. tinctoria* (Plains coreopsis)
Cornus drummondii (Rough-leaf
dogwood)
*Corydalis aurea (Golden corydalis)
**C. curvisiliqua* (Curved-pod
corydalis)
**Coryphantha missouriensis* (Cory
cactus)
***Crataegus marshallii* (Parsley
hawthorn)
C. mollis (Downy hawthorn)
**C. sutherlandensis* (Sutherland's
hawthorn)
**C. texana* (Texas hawthorn)
**C. viridis* (Green hawthorn)
*Croton texensis (Texas croton)
**Cucurbita foetidissima* (Buffalo
gourd)
Cuscuta sp. (Dodder)
*Cynanchum laeve (Swallow-wort)
*Cynodon dactylon (Bermuda grass)
**Cyperus acuminatus* (Taper-leaf flat
sedge)

**C. croceus* (Baldwin flat sedge)
**C. haspan* (Sheathed flat sedge)
C. involucratus (Umbrella sedge)
**C. polystachyos* (Branched flat sedge)
**C. retrorsus* (Pine barren flat sedge)
C. strigosus (False nut-grass)
***C. surinamensis* (Tropical flat sedge)
**C. uniflorus* (*C. retroflexus*) (One-
flowered flat sedge)
*Dalea emarginata (Wedgeleaf prairie-
clover)
**D. phleoides* var. *microphylla* (Long-
bract prairie-clover)
**D. tenuis* (Slender prairie-clover)
Delphinium carolinianum (Blue
larkspur)
**Desmatodon plinthobius* (Moss)
**Desmodium canescens* (Hoary tick-
clover)
**D. glabellum* (Smooth tick-clover)
Dichondra carolinensis (Ponyfoot)
Dicliptera brachiata (False mint)
***Digitaria sanguinalis* (Hairy crab
grass)
Diospyros texana (Texas persimmon)
**Ditrichum pallidum* (Moss)
*Draba cuneifolia (Wedgeleaf draba)
**Echinochloa colona* (Jungle-rice)
**E. crus-galli* (Barnyard grass)
Ehretia anacua (Anaqua)
***Eichhornia crassipes* (Water-
hyacinth)
**Eleocharis geniculata* (Jointed spike
sedge)
**E. interstincta* (Septate spike sedge)
***E. palustris* (Large-spike spike sedge)
**E. tenuis* (Slender spike sedge)
**E. tortilis* (Twisted spike sedge)
Elephantopus carolinianus (Leafy
elephant's foot)
**Elymus canadensis* (Canada wild-rye)
***E. virginicus* (Virginia wild-rye)
**Entodon seductrix* (Moss)
**Epipactis gigantea* (Giant helleborine)

Equisetum laevigatum (Smooth horsetail)
*Eragrostis hypnoides (Teal love grass)
**E. intermedia (Plains love grass)
**E. lugens (Mourning love grass)
*E. pectinacea (Spreading love grass)
**E. secundiflora subsp. *oxylepis* (Red love grass)
*Erichtites hieracifolia (American burnweed)
*Erigeron philadelphicus (Philadelphia fleabane)
**E. strigosus (Prairie fleabane)
**Eryngium diffusum (Bushy eryngo)
Erythrina herbacea (Coral bean)
*Eucnide bartonioides (Yellow rock-nettle)
Eupatorium coelestinum (Mist flower)
*E. incarnatum (Pink boneset)
Euphorbia (*Agaloma*) *marginata* (Snow-on-the-mountain)
*E. (*Chamaesyce*) *albomarginata* (White-margined spurge)
E. (*C.*) *serpens* (Mat euphorbia)
*E. (*Poinsettia*) *cyathophora* (Wild poinsettia)
*E. (*P.*) *dentata* (Toothed spurge)
*E. (*Tithymalus*) *spathulata* (Warty spurge)
*Eustoma grandiflorum (Bluebells)
*Fimbristylis puberula (Hairy fimbristylis)
*F. vahlii (Vahl's fimbristylis)
**Fissidens bryoides (Moss)
**F. bushii (Moss)
*Fraxinus berlandieriana (Mexican ash)
F. pennsylvanica (Green ash)
*Fuirena simplex (Western umbrella sedge)
*Funaria americanum (Moss)
**Gaillardia fastigiata (Prairie gaillardia)
*G. suavis (Sweet gaillardia)
*Galium tinctorium (Dye bedstraw)

*Gamochaeta falcata (Falcate cudweed)
**G. purpurea (Purple cudweed)
*Gaura drummondii (Drummond's gaura)
*G. suffulta (Roadside gaura)
Geum canadense (White avens)
*Grindelia squarrosa (Tarweed)
Gutierrezia texana (Texas broomweed)
*Habenaria repens (Water spider orchid)
Habranthus tubispathus (Copper-lily)
**Hedyotis nigricans (Star-violet)
*Helenium microcephalum (Small-head sneezeweed)
**Helianthemum carolinianum (Carolina sunrose)
*Helianthus angustifolius (Swamp sunflower)
*H. argophyllus (Silver-leaf sunflower)
*H. grosseserratus (Sawtooth sunflower)
Heliotropium indicum (Turnsole)
**Herbertia lahue (Herbertia)
Hibiscus laevis (Halberd-leaf rose-mallow)
*Hordeum pusillum (Little barley)
*Hydrocotyle ranunculoides (Floating water-pennywort)
H. umbellata (Umbrella water-pennywort)
H. verticillata (Whorled water-pennywort)
*Hygroamblystegium tenax (Moss)
**Hymenocallis caroliniana (Carolina spider-lily)
*Ibervillea lindheimeri (Lindheimer's globe-berry)
Ilex decidua (Possumhaw)
I. vomitoria (Yaupon)
*Ipomoea sagittata (Saltmarsh morning glory)
Iris hexagona var. *flexicaulis* (Purple fleur-de-lis, Dixie iris)

***I. pseudacorus* (Yellow flag)
***Isopterygium tenerum* (Moss)
*Iva angustifolia (Narrow-leaf
  sumpweed)
**I. annua* (Marsh-elder)
**Juncus diffusissimus* (Slimpod rush)
J. effusus (Soft rush)
***J. interior* (Inland rush)
**J. tenuis* (Path rush)
***J. validus* (Roundhead rush)
Juniperus virginiana (Eastern red-
  cedar)
***Lactuca canadensis* (Wild lettuce)
***L. floridana* (Florida lettuce)
Lamium amplexicaule (Dead-nettle)
**Lappula occidentalis* (Hairy
  stickweed)
**Lathyrus pusillus* (Low pea-vine)
**Leersia oryzoides* (Rice cutgrass)
*Leskea australis (Moss)
**Lespedeza violacea* (Violet bush-
  clover)
**Lesquerella grandiflora* (Big-flower
  bladderpod)
**L. recurvata* (Slender bladderpod)
**Leucobryum albidum* (Moss)
**Limnodea arkansana* (Ozark grass)
Lindernia dubia (Clasping false
  pimpernel)
Lindheimera texana (Texas yellow-
  star)
**Lipocarpha micrantha* (Common
  hemicarpha)
Lippia lanceolata (Northern frog-fruit)
Lobelia cardinalis (Cardinal flower)
**Lolium perenne (Perennial ryegrass)
**Lonicera albiflora* (White
  honeysuckle)
**Ludwigia alternifolia* (Bushy seed-
  box)
**L. glandulosa* (Cylindric-fruited
  ludwigia)
**L. leptocarpa* (Angle-stem water-
  primrose)

**L. peploides* (Creeping water-
  primrose)
Lupinus subcarnosus (Sandyland
  bluebonnet)
***L. texensis* (Texas bluebonnet)
Lycium carolinianum (Carolina
  wolfberry)
**Lygodesmia texana* (Skeleton plant)
*Lyonia mariana (Staggerbush)
*Lythrum alatum var. *lanceolatum*
  (Lance-leaf loosestrife)
**Maclura pomifera (Osage-orange)
**Malva rotundifolia* (Common mallow)
***Malvastrum aurantiacum* (Wright's
  false mallow)
M. coromandelianum (Three-lobe false
  mallow)
Malvaviscus arboreus (Texas-mallow,
  Turk's cap)
**Manfreda maculosa* (Spotted
  American aloe)
*Mannia fragrans (Liverwort)
**Marsilea macropoda* (Water-clover,
  a fern)
**M. vestita* (Hairy pepperwort, a fern)
**Matelea brevicoronata* (Short-
  crowned milkvine)
*Medicago arabica (Spotted bur-
  clover)
***M. minima* (Small bur-clover)
**M. sativa* (Alfalfa)
**Melampodium cinereum* (Hoary
  blackfoot-daisy)
Melia azedarach (Chinaberry)
*Melilotus albus (White sweet-clover)
***M. indicus* (Annual yellow sweet-
  clover)
***M. officinalis* (Yellow sweet-clover)
Melothria pendula (Drooping
  melonette)
*Mentzelia nuda (Bractless mentzelia)
Mikania scandens (Climbing boneset)
Mitreola (*Cynoctonum mitreola*)
  petiolata (Lax hornpod)

Modiola caroliniana (Carolina
modiola)
*Mollugo verticillata (Indian
chickweed)
**Monarda citriodora (Horsemint)
*M. fistulosa (Wild bergamot)
*M. lindheimeri (Lindheimer's
beebalm)
**M. punctata (Spotted beebalm)
Morus rubra (Red mulberry)
**Muhlenbergia schreberi (Satin grass)
**Myosotis macrosperma (Forget-me-
not)
Myrica cerifera (Southern wax-myrtle)
*Myriophyllum heterophyllum (Parrot's
feather)
*Nama jamaicense (Fiddle-leaf nama)
*Nassella leucotricha (Texas winter
grass)
*Nemastylis geminiflora (Prairie
celestial-lily)
Nemophila phacelioides (Baby blue
eyes)
*Nemophylla aphylla (Small-flowered
nemophila)
**Nitella flexilis (Bassweed, a green
alga)
Nothoscordum bivalve (Crow-poison)
Nuphar advena (Spatter-dock)
Nuttallanthus texanus (Texas toad-
flax)
Oenothera speciosa (Showy evening-
primrose)
O. triloba (Three-lobed evening-
primrose)
**Oldenlandia uniflora (One-flowered
bluets)
Onosmodium bejariense (Bejar
marbleseed)
*Ophioglossum engelmannii
(Engelmann's adder's tongue, a fern)
Oplismenus hirtellus (Basketgrass)
Opuntia humifusa (Eastern prickly-
pear)
O. leptocaulis (Pencil cactus)

Osmunda cinnamomea (Cinnamon
fern)
*Oxalis lyonii (Yellow wood-sorrel)
**Palafoxia callosa (Small palafoxia)
*P. hookeriana (Showy palafoxia)
Pallavicinia lyelli (Liverwort)
*Panicum anceps (Beaked panic grass)
**P. (Dichanthelium) acuminatum var.
lindheimeri (Lindheimer's rosette
grass)
*P. (D.) dichotomum (Paired rosette
grass)
**P. (D.) nodatum (Rosette grass)
*P. (D.) oligosanthes (Few-flowered
rosette grass)
P. (D.) scoparium (Velvet rosette grass)
P. divergens (Variable rosette grass)
*P. gymnocarpon (Savannah panic
grass)
P. virgatum (Switchgrass)
**Parietaria floridana (Florida
pellitory)
*P. obtusa (Blunt pellitory)
*P. pensylvanica (Pennsylvania
pellitory)
*Parthenium hysterophorus (False
ragweed)
*Parthenocissus heptaphylla (Seven-
leaf creeper)
P. quinquefolia (Virginia-creeper)
*Paspalum distichum (Knot grass)
*P. floridanum (Florida paspalum)
**Pellia epiphylla (Liverwort)
*Phacelia congesta (Woolly blue-curls)
*P. hirsuta (Hairy phacelia)
*P. strictiflora (Prairie phacelia)
Phaeoceros laevis (Hornwort)
*Phalaris angusta (Timothy canary
grass)
**P. canariensis (Reed canary grass)
*Phascum cuspidatum (Moss)
*Philonotis longiseta (Moss)
Phlox drummondii (Drummond's
phlox)

P. pilosa subsp. *latisepala* (*P. villosissima*) (Big-sepal phlox)

Phoradendron (*serotinum* var.) *tomentosum* (Mistletoe)

Phragmites australis (Common reed)

*Physalis cinerascens (Ground-cherry)

*P. heterophylla (Clammy ground-cherry)

*P. mollis (Field ground-cherry)

P. pubescens (Downy ground-cherry)

*Physcomitrium acuminatum (Moss)

*Physostegia intermedia (Intermediate lion's heart)

Phytolacca americana (Pokeweed)

*Plantago heterophylla (Many-seed plantain)

P. rhodosperma (Red-seed plantain)

Platanus occidentalis (American sycamore)

*Pluchea camphorata (Camphor weed)

P. odorata (*P. purpurascens*) (Purple marsh-fleabane)

*Poa annua (Annual bluegrass)

*Pogonia ophioglossoides (Rose pogonia)

**Polycarpon tetraphyllum (Four-leaf many-seed)

Polygonum hydropiperoides (Swamp smartweed)

*P. pensylvanica (Pennsylvania smartweed)

*P. sagittatum (Tearvine)

**Polypogon monspeliensis (Rabbit's-foot grass)

Populus deltoides var. *deltoides* (Eastern cottonwood)

*Porella pinnata (Liverwort)

*Potamogeton illinoensis (Shining pondweed)

**P. pusillus (Thread-leaf pondweed)

Prosopis glandulosa (Honey mesquite)

*Prunus mexicana (Mexican plum)

*P. texana (Texas almond)

Pteridium aquilinum var. *pseudocaudatum* (Bracken)

Ptilimnium capillaceum (Thread-leaf mock bishop's weed)

**P. nuttallii (Nuttall's mock bishop's weed)

**Pyrrhopappus carolinianus (Carolina false dandelion)

*P. grandiflorus (Tuberous false dandelion)

P. pauciflorus (Texas false dandelion)

*Quercus alba (White oak)

*Q. buckleyi (*texana*) (Texas oak)

*Q. falcata (Southern red oak)

Q. macrocarpa (Bur oak)

Q. shumardii (Shumard oak)

Q. sinuata (Bastard oak)

Ranunculus fascicularis (Tufted buttercup)

R. macranthus (Big buttercup)

R. pusillus (Weak buttercup)

**Ratibida columnifera (Mexican hat)

Rhexia mariana (Maryland meadow beauty)

**Rhus glabra (Smooth sumac)

*Rhynchosia americana (American snout-bean)

*Rhynchospora colorata (White-top umbrella-grass)

*R. glomerata (Cluster beak-rush)

Riccia fluitans (Liverwort)

*Ricinus communis (Castor-bean)

Rivina humilis (Pigeon berry)

*Rorippa nasturtium-aquaticum (Watercress)

Rubus trivialis (Southern dewberry)

**Rudbeckia hirta (Black-eyed Susan)

*Rumex chrysocarpus (Amamastla)

**R. hastatulus (Heart sorrel)

**R. pulcher (Fiddle dock)

**Sabal mexicana (Texas palmetto)

S. minor (Dwarf palmetto)

Sabatia campestris (Texas star)

*Saccharum giganteum (Sugarcane plume grass)

Sacciolepis striata (American cupscale grass)

Sagittaria graminea (Grass-leaf
arrowhead)
S. lancifolia (Scythe-fruit arrowhead)
*S. *montevidensis* (Long-lobed
arrowhead)
S. platyphylla (Delta arrowhead)
Salix nigra (Black willow)
Salvia farinacea (Mealy blue sage)
*S. *roemeriana* (Cedar sage)
Sambucus canadensis (Elderberry)
Samolus ebracteatus var. *alyssoides*
(Coastal brookweed)
*S. *parviflorus* (Thin-leaf brookweed)
Sanicula canadensis (Canada sanicle)
S. odorata (Cluster sanicle)
Sanvitalia ocymoides (Yellow
sanvitalia)
Sapindus saponaria var. *drummondii*
(Western soapberry)
Sassafras albidum (Sassafras)
Saururus cernuus (Lizard's tail)
Schoenocaulon drummondii
(Drummond's sabadilla)
Scirpus americanus (*Schoenoplectus
pungens*) (Sword-grass)
*S. *cyperinus* (Woolly-grass bulrush)
Sclerocarpus uniserialis (Mexican
bone-bract)
Senecio glabellus (Butterweed)
*S. *obovatus* (Golden groundsel)
*S. *tampicanus* (Groundsel)
Senna occidentalis (Coffee senna)
Sesbania drummondii (Drummond
rattlebox)
S. vesicaria (Bladderpod)
Setaria parviflora (Knot-root bristle
grass)
*S. *viridis* (Green foxtail)
Sida filicaulis (Spreading sida)
Sideroxylon (*Bumelia*) *lanuginosum*
(Ironwood)
Silphium radula (Rough-stem
rosinweed)
Silybum marianum (Milk thistle)

**Simsia calva* (Awnless bush-
sunflower)
Sinapsis alba (White mustard)
Sisymbrium officinale (Hedge-
mustard)
Sisyrinchium angustifolium (Bermuda
blue-eyed-grass)
**S. *chilense* (Sword-leaf blue-eyed-
grass)
S. pruinosum (Dotted blue-eyed-grass)
**Sium suave* (Water-parsnip)
[according to Turner et al. 2003a,
known in Texas only from the
Ottine Wetlands]
Smallanthus uvedalia (Bear's foot)
Smilax bona-nox (Fiddle-leaf
greenbrier)
*S. *glauca* (Catbrier)
*S. *renifolia* (Kidney-leaf greenbrier)
**S. *rotundifolia* (Bullbrier)
Solanum americanum (American
nightshade)
S. rostratum (Buffalo bur)
Solidago arguta var. *boottii* (Boott's
goldenrod)
*S. *gigantea* (Giant goldenrod)
S. rugosa (Rough-leaf goldenrod)
S. sempervirens (Seaside goldenrod)
Sophora secundiflora (Texas
mountain-laurel)
**Sorghum halapense* (Johnson grass)
Spartina spartinae (Gulf cordgrass)
**Spermolepis inermis* (Naked
scaleseed)
Sphagnum imbricatum (Peat moss)
*S. *lescurii* (Peat moss)
*S. *recurvum* (Peat moss)
Sphenopholis obtusata (Prairie
wedgescale)
**Spigelia loganioides* (*S. texana* in
part) (Florida pinkroot (Texas
pinkroot in part))
**Spirodela polyrhiza* (Duckmeat)
Sporobolus sp. (Dropseed)
Tauschia texana (Texas umbrellawort)

Taxodium distichum (Bald-cypress)
**Telaranea longifolia* (Liverwort)
**Tephrosia lindheimeri* (Lindheimer's
  hoary-pea)
**Teucrium canadense* (Wood
  germander)
**Thalia dealbata* (Powdery thalia)
Thelypteris kunthii (Southern shield
  fern)
**Tilia americana* var. *caroliniana*
  (Carolina basswood)
Tillandsia recurvata (Ball-moss)
T. usneoides (Spanish-moss)
Tinantia anomala (False dayflower)
***Torilis arvensis* (Field hedge-parsley)
***T. nodosa* (Knotted hedge-parsley)
Toxicodendron radicans (Poison-ivy)
Tradescantia gigantea (Giant
  spiderwort)
***Tribulus terrestris* (Puncture vine)
**Trifolium polymorphum* (Peanut
  clover)
**T. pratense* (Red clover)
***Typha domingensis* (Southern tule)
T. latifolia (Broad-leaf cattail)
Ulmus alata (Winged elm)
U. americana (American elm)
U. crassifolia (Cedar elm)
**Urochloa ciliatissima* (Fringed signal
  grass)
**Urtica chamaedryoides* (Low spring
  nettle)
**Valerianella amarella* (Hairy corn-
  salad)
**V. stenocarpa* (Narrow-cell corn-
  salad)

**Verbena plicata* (Fan-leaf vervain)
***V. scabra* (Harsh vervain)
**V. urticifolia* (Nettle-leaf vervain)
V. xutha (Gulf vervain)
Verbesina virginica (Frostweed)
**Veronica peregrina* (Necklace weed)
**Viburnum nudum* (Possumhaw
  viburnum)
V. rufidulum (Rusty blackhaw
  viburnum)
**Vicia minutiflora* (Pygmy-flowered
  vetch)
**Viguiera dentata* (Plateau goldeneye)
Viola missouriensis (Missouri violet)
**V. primulifolia* (Primrose-leaf violet)
Vitis cinerea (Sweet grape)
***V. mustangensis* (Mustang grape)
***V. vulpina* (Winter grape)
***Wolffia brasiliensis* (Dotted
  watermeal)
***W. columbiana* (Common
  watermeal)
Woodwardia areolata (Narrow-leaf
  chain fern)
W. virginica (Virginia chain fern)
**Yucca arkansana* (Arkansas yucca)
**Y. treculeana* (Spanish dagger)
Xanthium strumarium (Cocklebur)
**Xyris caroliniana* (Carolina yellow-
  eyed-grass)
**X. laxifolia* var. *iridifolia* (Iris-leaf
  yellow-eyed-grass)
Zizaniopsis miliacea (Giant cutgrass)
**Ziziphus obtusifolia* (Lotebush)

Glossary

alluvial originating as a deposit left behind by flowing water. Mud deposited on high ground by the flooding San Marcos River is alluvial in origin.

artesian well a well that flows because of natural pressure exerted by the weight of overlying water.

bog a nutrient-poor peatland that, according to strict usage, receives all of its water from precipitation. This condition arises when peat piles to such a height that the surface no longer has a connection with groundwater or surface water. In one view, bog peat has the further requirement that it must derive primarily from sphagnum mosses. By these criteria there are no true bogs in the Ottine Wetlands.

dioecious male and female flowers separate and on different plants.

fen a peatland that receives surface water and groundwater and that is nutrient rich as opposed to the atmospherically watered, nutrient-poor bog.

floodplain the region on either side of a river that is susceptible to being covered by water during floods.

gametophyte the individual in the life cycle of a plant that produces sperm or eggs.

glomerule a small, dense cluster or head.

groundwater water originating, most recently at least, from underground. This is the water of seeps, springs, and wells.

lagoon either a shallow pond communicating with a larger body of water, or a shallow artificial pond. Combining the two definitions and specifying the San Marcos River as the larger body of water that communicates via flooding, the modified, ephemeral ponds of Palmetto State Park may be considered lagoons.

marsh a wetland dominated by herbaceous vegetation that grows during at least part of the year from water-covered or water-saturated ground.

mericarps the paired, seed-containing units of a mature flower in the parsley family (Apiaceae).

mesic hammock relatively high ground within the wetland where soil rich in decomposing vegetation occurs.

mud boil a point in a wetland where gas and/or liquid rises to the surface and escapes from mud in the form of bubbles.

oogonia seedlike, egg-producing structures in algae.

oxbow lake a body of water left behind in the old river bed when a river changes its course.

panicle a multiply branched flower cluster

maturing from the bottom upward—a compound raceme.

peat an accumulation of partly decomposed plant material.

peatland land covered by peat but especially those vast tracts of peat-covered land unique to more northern states and countries.

perfect a flower with both male (stamens) and female parts (pistils).

pH a measure of acidity. The lower the pH, the greater the acidity of the water.

pinnate a featherlike compound leaf with leaflets arranged in two opposite rows like the vanes of a feather.

quaking a jellylike quivering of certain water-saturated, thickly matted peatlands. Quaking may be induced by jumping up and down in one spot or merely by walking.

raceme an inflorescence with stalked flowers arising from an unbranched, central axis, and maturing from the bottom upward.

seep a place where water reaches the surface from underground while flowing at a slow rate.

slough a muddy inlet or creek.

sporophyte the individual in the life cycle of a plant that produces spores.

spring a fast-flowing seep.

streamlet groundwater flowing out of the ground at a rate and volume between that of a seep and a spring. We found these flowing cold even in summer in North Soefje Marsh.

surface water the runoff from rain, and, according to one view, the water of rivers and creeks as well.

swamp a wetland dominated by trees that grow during at least part of the year from water-covered ground.

wetland a general term meaning land that, during at least part of the year, is wet enough to support the growth of water-loving vegetation. Water may cover the ground or merely saturate it. Bogs, marshes, peatlands, and swamps are all varieties of wetland.

wet meadow a synonym of *fen* but certainly more descriptive.

Bibliography

Bailey, V. 1905. *Biological survey of Texas.* North American Fauna No. 25. Washington, D.C.: U.S. Department of Agriculture Biological Survey.

Banks, S., H. Hollister, and C. Llewellin. 2004. This land is your land. *Texas Monthly,* March, 105–19.

Barlow, C. 2000. *The ghosts of evolution.* New York: Basic Books.

Blair, W. F. 1950. The biotic provinces of Texas. *Tex. J. Sci.* 2:93–117.

Bogusch, E. R. 1928. Composition and seasonal aspects of the Gonzales County marsh associes. Master's thesis, University of Texas at Austin.

———. 1930. Trend of succession in a southern bog. *Ill. Acad. Sci., Papers in Biology and Agriculture* 23:285–97.

Bryant, V. M., Jr. 1977. A 16,000 year pollen record of vegetational change in Central Texas. *Palynology* 1:143–56.

Bullard, F. M. 1935. Geological background. In Palmetto State Park, Texas Board, 18–19.

Bureau of Business Research and Natural Fibers Information Center. 1987. *The climates of Texas counties.* College Station: University of Texas at Austin in cooperation with the Office of State Climatologist, Texas A&M University.

Bureau of Economic Geology. 1974. *Geologic atlas of Texas.* Austin and Seguin sheets. Austin: University of Texas.

Chelf, C. 1941. *Peat bogs in Gonzales County with notes on other bogs.* Circular No. 34. Austin: Bureau of Economic Geology, University of Texas.

Cholewa, A. F., and D. M. Henderson 2002. *Sisyrinchium.* In Flora of North America Editorial Committee, 2002b, 351–71.

Clay, K. 1993. Size-dependent gender change in green dragon (*Arisaema dracontium;* Araceae). *Am. J. Bot.* 80:769–77.

Conard H. S., and P. L. Redfearn, Jr. 1979. *How to know the mosses and liverworts.* 2nd ed. Dubuque, Iowa: Wm. C. Brown.

Cope, E. D. 1880. On the zoological position of Texas. *Bull. U.S. Nat. Mus.* 17:1–51.

Corner, E. J. H. 1966. *The natural history of palms.* Berkeley: University of California Press.

Correll, D. S., and H. B. Correll. 1972. *Aquatic and wetland plants of southwestern United States.* Washington, D.C.: Environmental Protection Agency.

Correll, D. S., and M. C. Johnston. 1970. *Manual of the vascular plants of Texas.* Renner: Texas Research Foundation.

Crum, H. A., and L. E. Anderson. 1981. *Mosses of eastern North America.* Vols. 1, 2. New York: Columbia University Press.

Cumley, R. W. 1931. *A geologic section across Caldwell County, Texas.* Master's thesis, University of Texas at Austin.

Delcourt, P. A., and H. R. Delcourt. 1981. Vegetation maps for eastern North America: 40,000 yr. B.P. to the present. In *Geobotany II,* ed. R. C. Romans, 123–65. New York: Plenum Press.

Diggs, G. M., Jr., B. L. Lipscomb, and R. J. O'Kennon. 1999. *Shinner & Mahler's illustrated flora of north central Texas.* Sida, Botanical Miscellany 16. Fort Worth: Botanical Res. Inst. Texas.

Diggs, G. M., Jr., B. L. Lipscomb, M. D. Reed, and R. J. O'Kennon. 2006. *Illustrated flora of East Texas.* Vol. 1, *Introduction, pteridophytes, gymnosperms, and monocotyledons.* Sida, Botanical Miscellany 26. Fort Worth: Botanical Res. Inst. Texas.

Ellison, M. L. 1964. The liverwort flora of Texas. Ph.D. diss., University of Kansas.

Engel, J. J., and G. L. Smith Merrill. 2004. *Austral Hepaticae. 35. A taxonomic and phylogenetic study of* Telaranea *(Lepidoziaceae), with a monograph of the genus in temperate Australasia and commentary on extra-Australasian taxa.* Fieldiana. Botany New Series, No. 44. Chicago: Field Museum of Natural History, Publication 1531.

Faden, R. B. 2000. *Tradescantia.* In Flora of North America Editorial Committee, 2000, 173–87.

Flora of North America Editorial Committee, eds. 1993a. *Flora of North America north of Mexico.* Vol. 1, *Introduction and vegetation.* New York: Oxford University Press.

———. 1993b. *Flora of North America north of Mexico.* Vol. 2, *Pteridiphytes and Gymnosperms.* New York: Oxford University Press.

———. 1997. *Flora of North America north*

of Mexico. Vol. 3, *Magnoliophyta: Magnoliidae and Hamamelidae.* New York: Oxford University Press.

———. 2000. *Flora of North America north of Mexico.* Vol. 22, *Magnoliophyta: Alismatidae, Arecidae, Commelinidae (in part), and Zingiberidae.* New York: Oxford University Press.

———. 2002a. *Flora of North America north of Mexico.* Vol. 23, *Magnoliophyta: Commelinidae (in part): Cyperaceae.* New York: Oxford University Press.

———. 2002b. *Flora of North America north of Mexico.* Vol. 26, *Magnoliophyta: Liliidae: Liliales and Orchidales.* New York: Oxford University Press.

———. 2005. *Flora of North America north of Mexico.* Vol. 5, *Magnoliophyta: Caryophyllidae, pt. 2.* New York: Oxford University Press.

Foster, J. H. 1917. The spread of timbered areas in Texas. *J. Forestry* 15:442–45.

Godfrey, R. K., and J. W. Wooten. 1979. *Aquatic and wetland plants of southeastern United States.* Vol. I, *Monocotyledons.* Athens: University of Georgia Press.

———. 1981. *Aquatic and wetland plants of southeastern United States.* Vol. II, *Dicotyledons.* Athens: University of Georgia Press.

Graham, A. 1958. *Pollen studies of some Texas peat deposits.* Master's thesis, University of Texas at Austin.

Graham, A., and C. Heimsch. 1960. Pollen studies of some Texas peat deposits. *Ecology* 41:751–63.

Hall, S. A., and S. Valastro, Jr. 1995. Grassland vegetation in the southern Great Plains during the last glacial maximum. *Quaternary Res.* 44:237–45.

Hartigan, P., and G. Lasley, comp. 1987. *Birds of Palmetto State Park: A field checklist.* Austin: Resource Management Section, Texas Parks and Wildlife Department.

Henderson, A., G. Galeano, and R. Bernal.

1995. *Field guide to the palms of the Americas.* Princeton, N.J.: Princeton University Press.

Henderson, N. C. 2002. *Iris.* In Flora of North America Editorial Committee, 2002b, 371–95.

Henrickson, J. 1996. Notes on *Spigelia* (Loganaciae). *Sida* 17:89–103.

Hildebrand, E. F. 1935. History of Palmetto State Park. In Palmetto State Park, Texas Board, 2–4.

Holm, L., J. Doll, E. Holm, J. Pancho, and J. Herberger. 1997. *World weeds. Natural histories and distribution.* New York: Wiley.

Horn, C. N. 2002. Pontederiaceae. In Flora of North America Editorial Committee, 2002b, 37–46.

King, E. A., Jr. 1961. *Geology of northwestern Gonzales County.* Master's thesis, University of Texas at Austin.

Kirn, A. J. 1935. Birds of the Ottine area. In Palmetto State Park, Texas Board, 12–14.

Kral, R. 1955. A floristic comparison of two hillside bog localities in northeastern Texas. *Field & Laboratory* 23:47–69.

Kral, R., and P. E. Bostick. 1969. The genus *Rhexia* (Melastomataceae). *Sida* 3:387–440.

Kushlan, J. A. 1993. Freshwater wetlands. In *Wetlands: Guide to science, law, and technology,* ed. M. S. Dennison and J. F. Berry, 74–127. Park Ridge, N.J.: Noyes Publications.

Larson, D. A., V. M. Bryant, and T. S. Patty. 1972. Pollen analysis of a Central Texas bog. *Am. Midl. Nat.* 88:358–67.

Lewis, W. M., Jr. 2001. *Wetlands explained.* Oxford: Oxford University Press.

Lockett, L. 2003. The native presence of *Sabal mexicana* (*Sabal texana*) north of the lower Rio Grande Valley. *NPSOT News* 21 (3): 1, 6–10.

Lodwick, L. N. 1975. *Net aerial primary productivity of three East Texas peat bogs.* Master's thesis, Baylor University, Waco, Texas.

Lodwick, L. N., and J. A. Snider. 1980. The distribution of *Sphagnum* taxa in Texas. *Bryologist* 83:214–18.

MacRoberts, B. R., and M. H. MacRoberts. 1998. Floristics of muck bogs in east central Texas. *Phytologia* 85:61–73.

MacRoberts, M. H., and B. R. MacRoberts. 2001. *Bog communities of the West Gulf Coastal Plain: A profile.* Bog Research Papers in Botany & Ecology, No. 1. Shreveport, La.: Bog Research.

Mahler, W. F. 1980. *The mosses of Texas. A manual of the moss flora with sketches.* Published by the author without copyright.

Martin, A. C., H. S. Zim, and A. L. Nelson. 1961. *American wildlife and plants. A guide to wildlife food habits: The use of trees, shrubs, weeds, and herbs by birds and mammals of the United States.* New York: Dover Publications.

Maxwell, R. A. 1970. Palmetto State Park. In *Geologic and historic guide to state parks of Texas,* 150–53. Bureau of Economic Geology Guidebook No. 10. Austin: University of Texas.

McAllister, F., P. Y. Hogland, and E. Whitehouse. 1930. Catalogue of Texas Hepaticae, with notes on the habitat and distribution. *J. Tex. Acad. Sci.* 15:39–58; pls. III, IV.

McWilliams, E. 1992. Chronology of the natural range expansion of *Tillandsia recurvata* (Bromeliaceae) in Texas. *Sida* 15:343–46.

Meanley, B. 1972. *Swamps, river bottoms and canebrakes.* Barre, Mass.: Barre Publishers.

Miglarese, J. V., and P. A. Sandifer, eds. 1982. *An ecological characterization of South Carolina wetland impoundments.* South Carolina Marine Resources Center Technical Report No. 51. Charleston: Marine Resources Research Institute, South Carolina Wildlife and Marine Resources Department.

Mitsch, W. J., and J. G. Gosselink. 2000. *Wetlands.* 3rd ed. New York: Wiley.

Mohlenbrock, R. H. 2002. Going with the flow: A Texas river winds through town and country. *Natural History,* February 2, 14–15.

Naczi, R. F. C., and C. T. Bryson. 2002. *Carex* sect. Griseae. In Flora of North America Editorial Committee, 2002a, 448–61.

Nixon, E. S. 1963. The role of soil in the distribution of certain Texas species. Ph.D. diss., University of Texas at Austin.

———. 1969. An edaphic study of two *Spartina spartinae* locations in Texas. *Tex. J. Sci.* 21:93–95.

Nixon, E. S., L. F. Chambless, and J. L. Malloy. 1973. Woody vegetation of a palmetto [*Sabal minor* (Jacq.) Pers.] area in East Texas. *Tex. J. Sci.* 24:535–41.

Nixon, K. C., and C. H. Muller. 1992. The taxonomic resurrection of *Quercus laceyi* Small (Fagaceae). *Sida* 15:57–69.

Palmetto State Park: Palmetto & river nature trails guide book. [undated park booklet for use with nature trails]

Palmetto State Park, Texas Board. 1935. *First Scientific Field Meet: Palmetto State Park, Ottine, Texas.* Gonzales (?): Palmetto State Park, Texas Board.

Parks, H. B. 1935a. Plant life of Ottine. In Palmetto State Park, Texas Board, 5–11.

———. 1935b. Amphibia and reptiles. In Palmetto State Park, Texas Board, 14–16.

———. 1935c. Butterflies of Ottine area. In Palmetto State Park, Texas Board, 16–18.

Patty, T. S. 1968. *Pollen analysis and chronology of a Central Texas peat bog.* Master's thesis, University of Texas at Austin.

Penfound, W. T. 1952. Southern swamps and marshes. *Bot. Rev.* 18:413–46.

Plummer, F. B. 1941. *Peat deposits in Texas.* University of Texas Bureau of Economic Geology, Mineral Resource Circular No. 13. Austin: University of Texas.

———. 1945. *Progress report on peat deposits in Texas.* University of Texas Bureau of Economic Geology, Mineral Resource Circular No. 36. Austin: University of Texas.

Potzger, J. E., and B. C. Tharp. 1943. Pollen record of Canadian spruce and fir from a Texas bog. *Science* 98:584–85.

———. 1947. Pollen profile from a Texas bog. *Ecology* 28:274–80.

———. 1954. Pollen study of two bogs in Texas. *Ecology* 35:462–66.

Ramsey, R. N., and N. P. Bade. 1977. *Soil survey of Guadalupe County, Texas.* College Station: Soil Conservation Service in cooperation with the Texas Agricultural Experiment Station.

Raun, G. G. 1958. *Vertebrates of a moist, relict area in Texas.* Master's thesis, University of Texas at Austin.

———. 1959. Terrestrial and aquatic vertebrates of a moist, relict area in Central Texas. *Tex. J. Sci.* 11:158–71.

Reznicek, A. A. 2002. *Carex* sect. Lupulinae. In Flora of North America Editorial Committee, 2002a, 511–14.

Reznicek, A. A., and B. A. Ford. 2002. *Carex* sect. Vesicariae. In Flora of North America Editorial Committee, 2002a, 501–11.

Rogers, C. D. 1999. *Birds of Palmetto State Park: A field checklist.* Austin: Natural Resources Program, Texas Parks and Wildlife Department.

Rowell, C. M., Jr. 1949. A preliminary report on the floral composition of a *Sphagnum* bog in Robertson County. *Tex. J. Sci.* 1:50–53.

Shearer, G. K. 1956. *Palmetto State Park: Ottine, Texas.* Austin: Texas State Parks Board.

Smeins, F. E., and D. D. Diamond. 1983. Remnant grasslands of the Fayette Prairie, Texas. *Am. Midl. Nat.* 110:1–13.

Soil Survey Staff. 1975. *Soil taxonomy. A basic system of soil classification for making and interpreting soil surveys.* Soil Conservation Service. USDA Agriculture Handbook no. 436. Washington, D.C.: USDA.

Stoneburner, A., and R. Wyatt. 1979. Three Big Thicket bryophytes new to Texas. *Bryologist* 82:491–93.

Studlar, S. M., and S. McAlister. 1994. Bryophyte communities at Boehler seeps, Oklahoma. (Abstract 32). *Am. J. Bot.* 81 (Suppl.): 12.

Stutzenbaker, C. D. 1999. *Aquatic and wetland plants of the western Gulf Coast.* Austin: Texas Parks and Wildlife Press.

Taber, S. W., and S. B. Fleenor. 2003. *Insects of the Texas Lost Pines.* Number 33, W. L. Moody, Jr., Natural History Series. College Station: Texas A&M University Press.

———. 2005. *Invertebrates of Central Texas wetlands.* Lubbock: Texas Tech University Press.

Talbot, S. S., and R. R. Ireland. 1982. Bryophytes new to the flora of Oklahoma. *Bryologist* 85:319.

Taylor, R. 1991. *The feral hog in Texas.* Texas Parks and Wildlife Department/ Federal Aid Report Series No. 28: A Contribution of Federal Aid (P-R) Project W-125-R. Austin: Texas Parks and Wildlife Department.

Texas Parks and Wildlife Department. 2000. Resource management plan for Palmetto State Park. Unpublished manuscript. Austin: Texas Parks and Wildlife Department, State Parks Division, Region 5.

Tharp, B. C. 1935. Unusual plants of the Ottine area. In Palmetto State Park, Texas Board, 4–5.

———. 1939. *The vegetation of Texas.* Texas Academy of Science Publications in Natural History no. 1. Houston: Anson Jones Press.

Thorne, R. F. 1993. Phytogeography. In Flora of North America Editorial Committee, 1993a, 132–53.

Tiner, R. W. 1999. *Wetland indicators:*

A guide to wetland identification, delineation, classification, and mapping. Boca Raton, Fla.: Lewis Publishers.

Turner, B. L. 1959. *The legumes of Texas.* Austin: University of Texas Press.

———. 1995. Synopsis of the genus *Onosmodium* (Boraginaceae). *Phytologia* 78:39–60.

Turner, B. L. , H. Nichols, G. Denny, and O. Doron. 2003a. *Atlas of the vascular plants of Texas.* Vol. 1, *Introduction, dicots.* Sida Botanical Miscellany No. 24. Fort Worth: Botanical Res. Inst. Texas.

———. 2003b. *Atlas of the vascular plants of Texas.* Vol. 2, *Ferns, gymnosperms and monocots.* Sida Botanical Miscellany No. 24. Fort Worth: Botanical Res. Inst. Texas.

U.S. Geological Survey. 1959 (photorevised 1988). Ottine, Tex. 7.5 minute Topographic Quadrangle Sheet. Washington, D.C.: U.S. Geological Survey.

Vitt, D. H. 1994. An overview of factors that influence the development of Canadian peatlands. *Mem. Entomol. Soc. Can.* 169:7–20.

———. 2000. Peatlands: Ecosystems dominated by bryophytes. In *Bryophyte biology,* ed. A. J. Shaw and B. Goffinet, 312–43. Cambridge: Cambridge University Press.

Wharton, C. H., W. M. Kitchens, and T. W. Sipe. 1982. *The ecology of bottomland hardwood swamps of the southeast: A community profile.* U.S. Fish and Wildlife Service OBS-81/37. Washington, D.C.: U.S. Fish and Wildlife Service.

Whitehouse, E. 1955. Additions to the hepatic flora of Texas. *Mitt. Thuring. Bot. Gesell. (Theodor Herzog Festschrift)* 1:231–35.

Whitehouse, E., and F. McAllister. 1954. The mosses of Texas. A catalogue with annotations. *Bryologist* 57:63–146.

Whittemore, A. T. 1997. *Ranunculus.*

In Flora of North America Editorial Committee, 1997, 88–135.

Williams, J. E., and W. A. Watson. 1978. Plants of the Ottine area including Palmetto State Park and the Rutledge Swamp. Unpublished manuscript intended as a Natural Areas Survey. From the Natural Resources Program archives, Texas Parks and Wildlife Department.

Wood, R. D., and K. Imahori. 1964. *A revision of the Characeae. Second Part. Iconograph of the Characeae.* Weinheim, Germany: J. Cramer; New York: Stechert-Hafner Service Agency.

———. 1965. *A revision of the Characeae. First Part. Monograph of the Characeae.* Weinheim, Germany: J. Cramer; New York: Stechert-Hafner Service Agency.

Zona, S. 1990. A monograph of *Sabal* (Arecaceae: Coryphoideae). *Aliso* 12:583–666.

Taxonomic Summary

	Number of Species
Algae	2
Liverworts	14
Hornworts	1
Mosses	31
Ferns and Allies	13
Conifers	2
Flowering Plants:	
Monocots	125
Dicots	336
Total species	524

Index

Scientific names and pages with relevant figures appear in *italics*.